KU-473-318

Larger than life

First published in 2001 by Miles Kelly Publishing,
Bardfield Centre, Great Bardfield, Essex CM7 4SL

Designer: Sarah Ponder
Project Manager: Paula Borton
Editorial Assistant: Isla MacCuish
Additional editorial assistance: Liz Tortice
Art Director: Clare Sleven

British Library Cataloguing-in-Publication Data
A catalogue record for this book is available from the British Library

ISBN 1-84236-020-5

24681097531

Visit us on the web:
www.mileskelly.net
Info@mileskelly.net

Printed in China

Larger than life

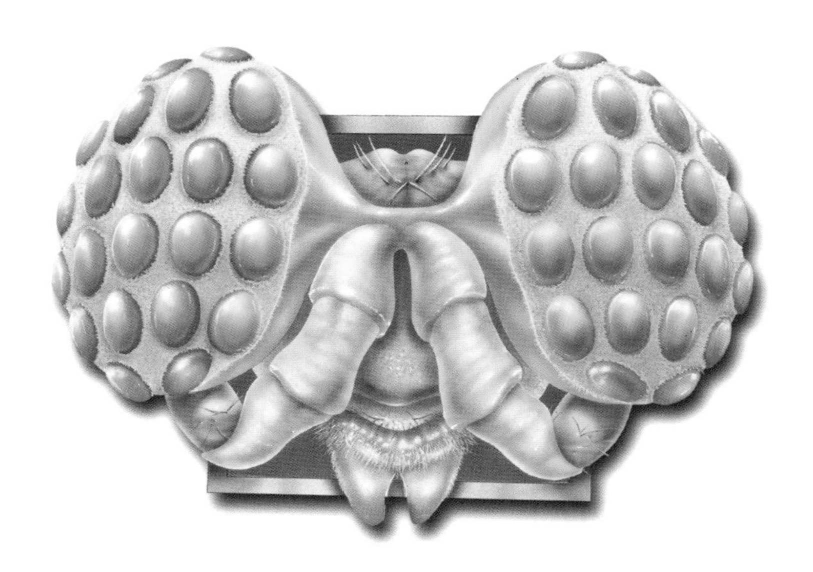

Steve Parker

Illustrated by Mike Saunders

Contents

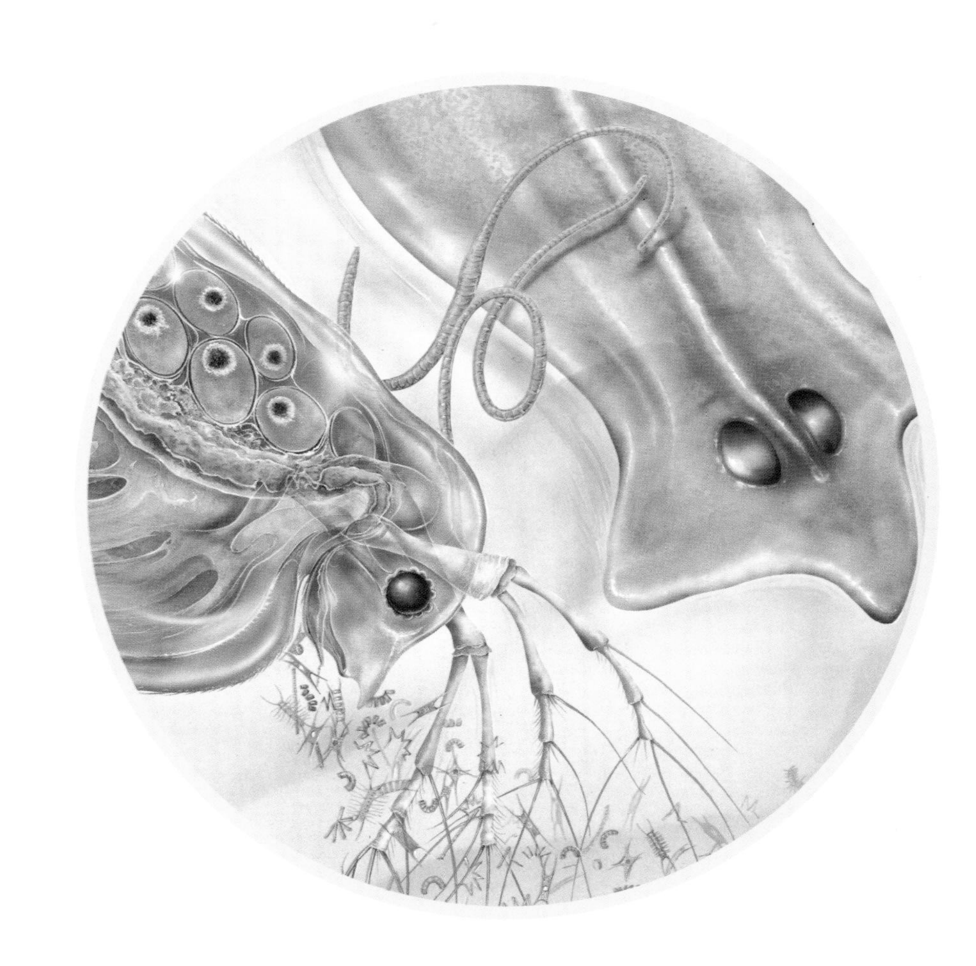

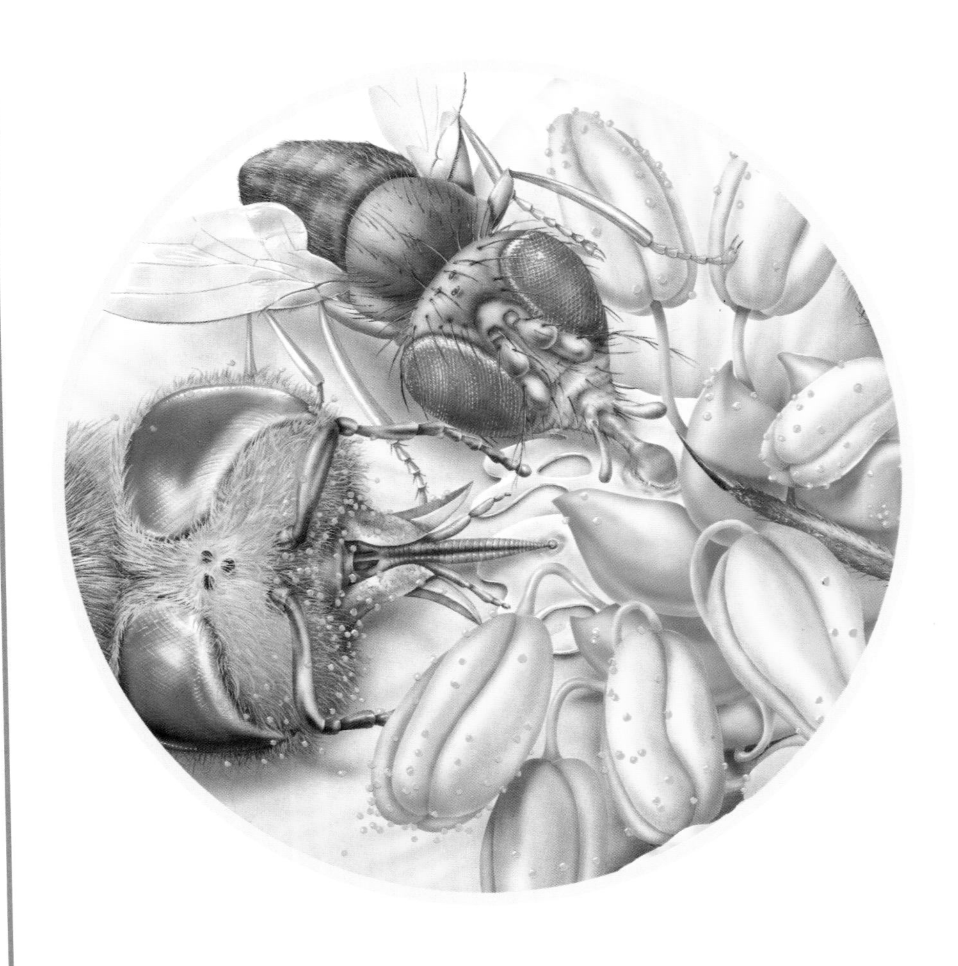

Soil

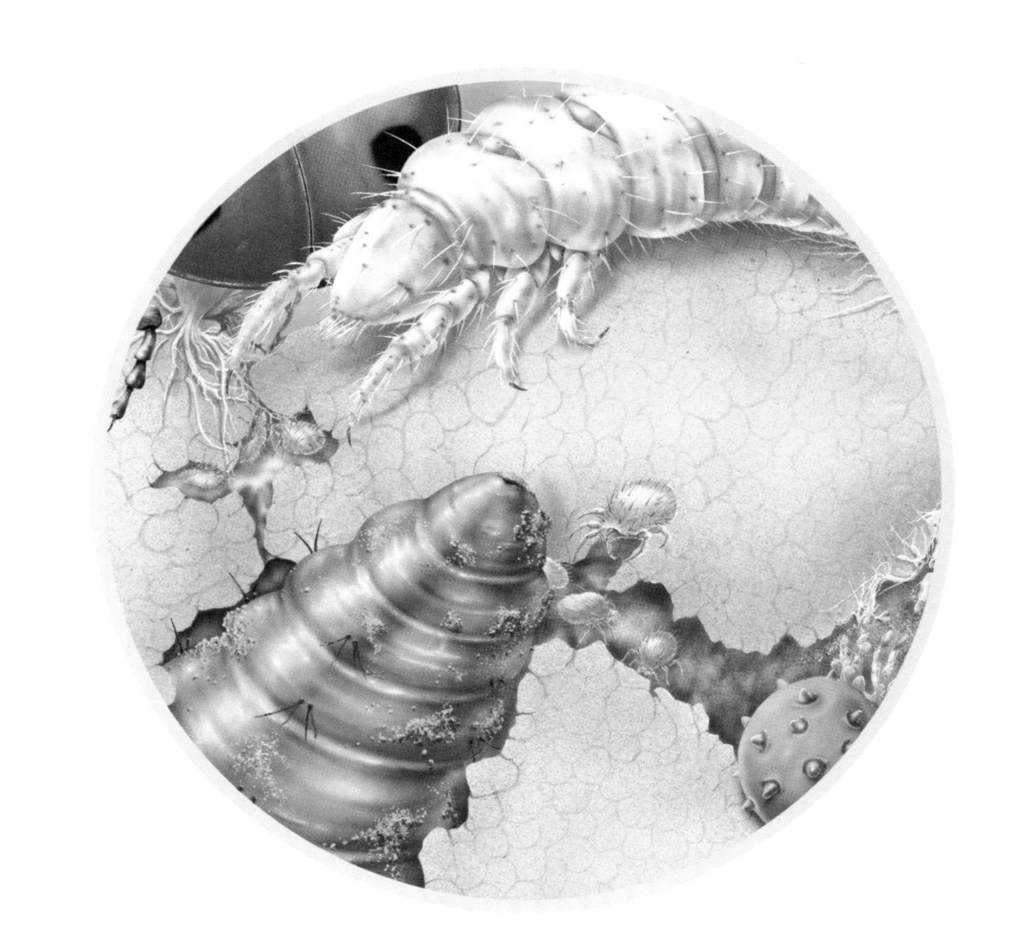

House and Human

Small, smaller, mini, micro, ultra, nano ...

We exist in a world geared to human scale. We make objects to suit the sizes of our bodies, such as chairs, beds, houses and cars. In nature, we are most familiar with animals and plants from the same size range. Horses are big. Mice are quite small. We can hardly see tinier creatures, such as houseflies and ants. They tend to be ignored, dismissed or swatted out of existence.

But these are only the limits of our eyes and our understanding. The real world goes much, much smaller. To see it, we must magnify it – make it larger than life. As we delve into this series of worlds within worlds, we find that creatures, bugs and blobs of jelly exist on almost every scale, almost everywhere. In drops of rain and sea water, in soil and mud, under bark and dead leaves, floating through the air, in the kitchen and carpet, even on our own skin and in blood – every patch of Earth is a micro-habitat for hordes of tiny wildlife.

The smallest creatures face similar survival tasks to "normal"-sized animals. They must find food, shelter and mates, and avoid predators and dangers. But their worlds are very different from ours. The physical aspects of the environment alter greatly with diminishing size.

A bigger view
Around 50 times life-size, creatures that just look like tiny dots to the naked eye, take on shape and form. The bark weevil is revealed with a long, pointed snout, feelers, eyes, and legs with tiny jointed sections.

x1 (Life size) x10 x100 x1000

Low power
At magnifications up to about 25–30, details of familiar animals such as flies begin to take shape. We can see their individual body hairs and the separate mosaic-like units of their eyes. But there's still a long way to go.

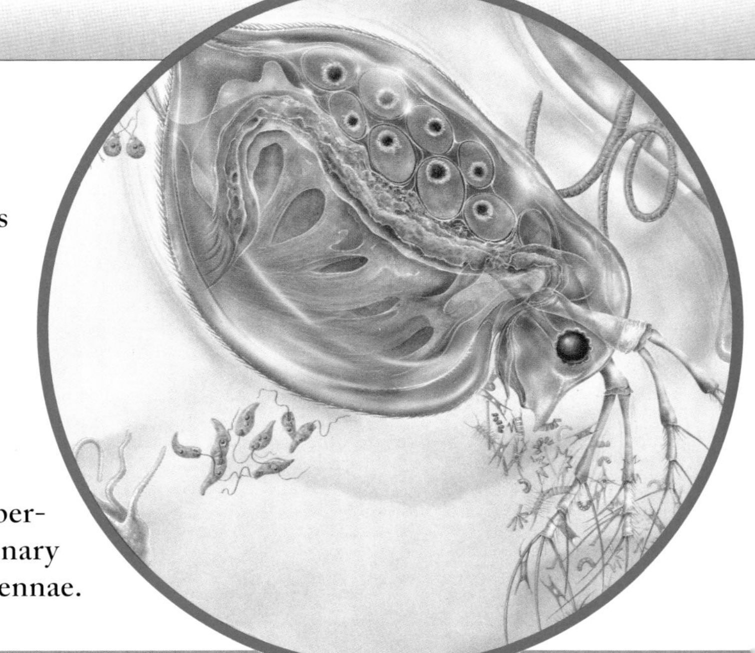

A different world
When enlargements reach the hundreds, the formerly invisible world becomes truly visible. The water flea, in reality as small as the dot on this i, becomes a super-beast with extraordinary multi-branched antennae.

What we feel as the merest breeze, is a howling gale that batters tiny gnats and moths. Water resists our wading and swimming to an extent, but it is like the thickest syrup to aquatic mini-beasts. We see a piece of tree bark as fairly smooth, but for the miniature tree-dwellers, it has more highs and lows than the Grand Canyon.

We also find it easier to understand and relate to creatures with familiar bodily features, such as eyes, a mouth and legs. But increasing magnification reveals life-forms which are far more difficult to recognize and comprehend, with no head, limbs, mouth, and in some cases, even no particular shape.

The following pages bring these mini- and micro-worlds vividly to life. They reveal the trials and problems of their inhabitants. And they lead to one overall, inescapable conclusion. Wherever we are, we are never alone.

x10,000

x100,000

x1000,000 (1 million)

Truly microscopic
Magnifications of thousands are required to visualize bacteria and similar micro-organisms (above). Many of these are responsible for diseases. They exist almost everywhere, unseen in air, soil, water, and on and in our bodies.

Limits of life
Some half a million times larger than life, the smallest organisms are viruses. At this level, individual molecules come into focus, for example, the protein subunits that make up the outer casing of each virus. The next stop would be small molecules and large atoms – but no life as we know it.

Float-aways

The clear, clean air all around us is far from that. We can just spot tiny flies and similar creatures as they flap with blurred wings into a gentle breeze – which for them is a howling hurricane. Smaller still and beyond our vision are flower pollen, spores, fragments discarded by animal fliers, and other bits of tiny debris that make up the invisible micro-blizzard.

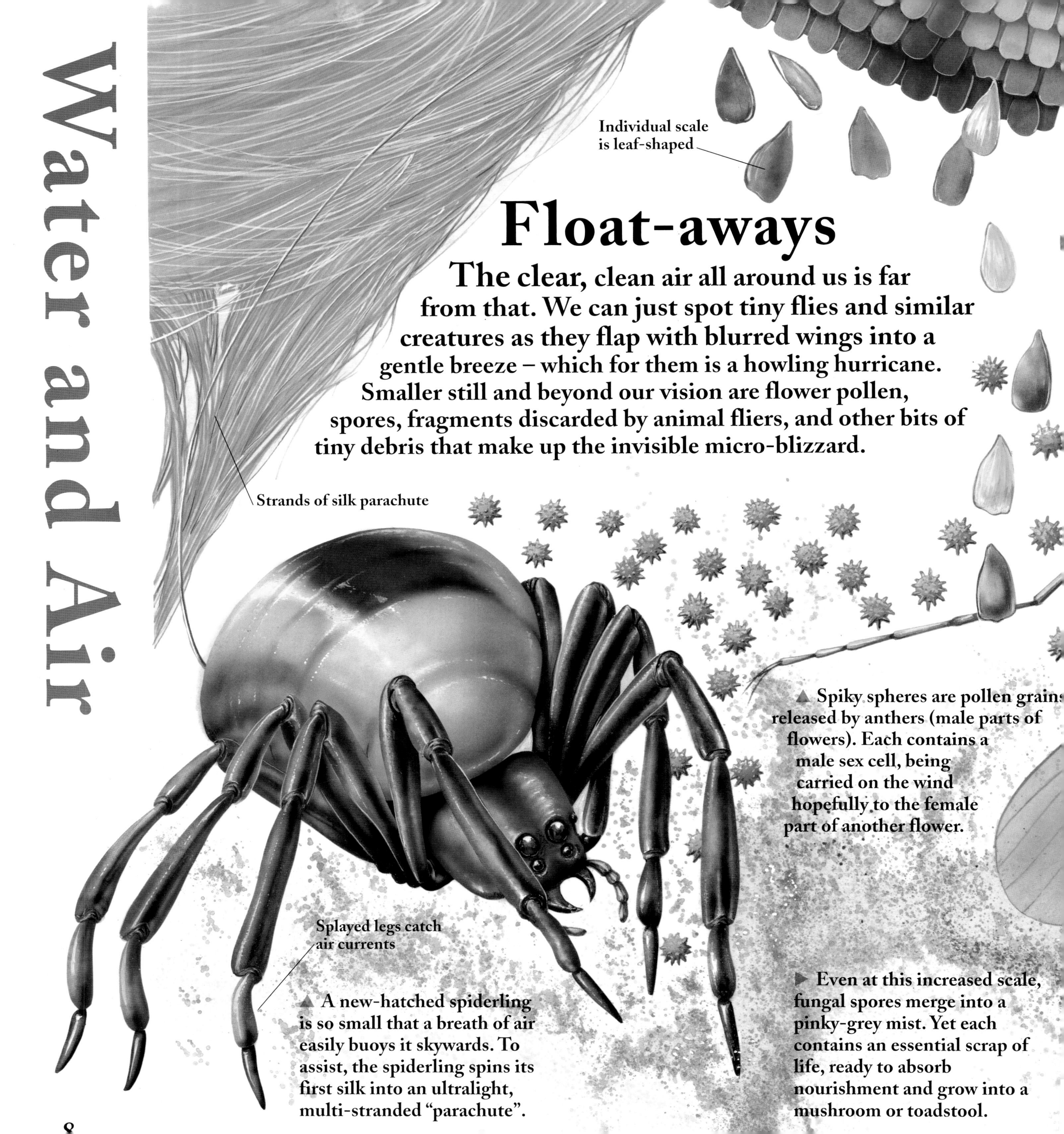

▲ Spiky spheres are pollen grains released by anthers (male parts of flowers). Each contains a male sex cell, being carried on the wind hopefully to the female part of another flower.

▲ A new-hatched spiderling is so small that a breath of air easily buoys it skywards. To assist, the spiderling spins its first silk into an ultralight, multi-stranded "parachute".

► Even at this increased scale, fungal spores merge into a pinky-grey mist. Yet each contains an essential scrap of life, ready to absorb nourishment and grow into a mushroom or toadstool.

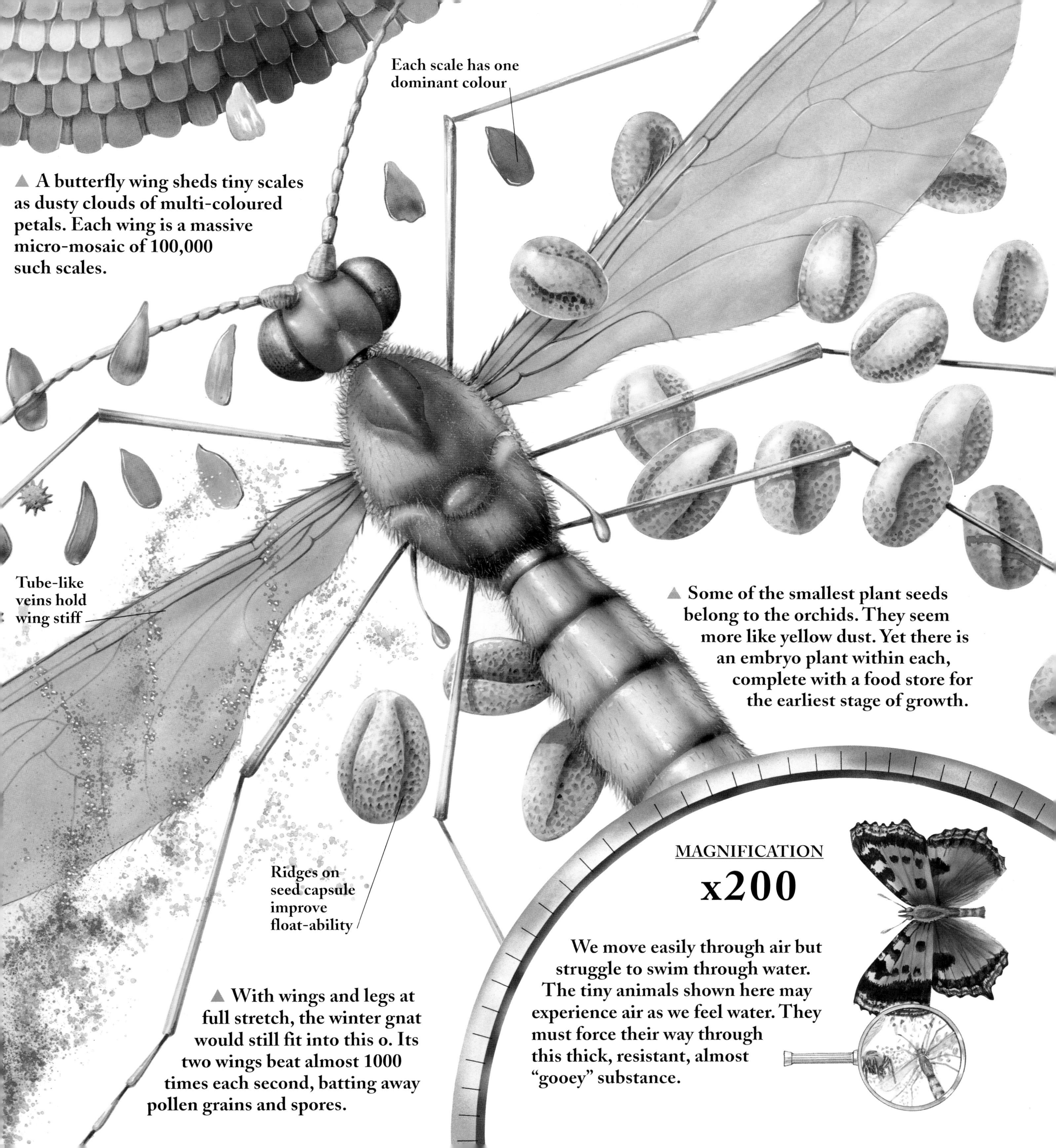
Each scale has one dominant colour
▲ A butterfly wing sheds tiny scales as dusty clouds of multi-coloured petals. Each wing is a massive micro-mosaic of 100,000 such scales.
Tube-like veins hold wing stiff
▲ Some of the smallest plant seeds belong to the orchids. They seem more like yellow dust. Yet there is an embryo plant within each, complete with a food store for the earliest stage of growth.
MAGNIFICATION
x200
Ridges on seed capsule improve float-ability
▲ With wings and legs at full stretch, the winter gnat would still fit into this o. Its two wings beat almost 1000 times each second, batting away pollen grains and spores.
We move easily through air but struggle to swim through water. The tiny animals shown here may experience air as we feel water. They must force their way through this thick, resistant, almost "gooey" substance.

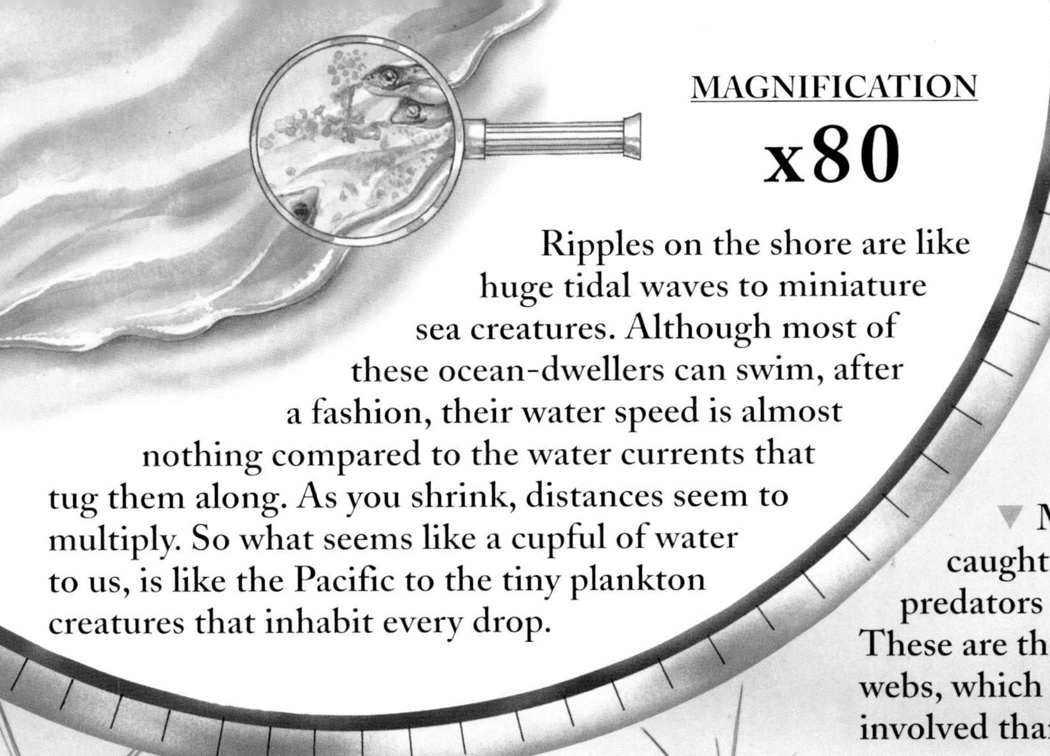

MAGNIFICATION

x80

Ripples on the shore are like huge tidal waves to miniature sea creatures. Although most of these ocean-dwellers can swim, after a fashion, their water speed is almost nothing compared to the water currents that tug them along. As you shrink, distances seem to multiply. So what seems like a cupful of water to us, is like the Pacific to the tiny plankton creatures that inhabit every drop.

▶ These feathery-tipped, multi-limbed blobs hardly resemble the rigid-shelled, cone-shaped barnacles which stick to seashore boulders. But that is what they will become. The larvae float to a suitable place, cement their head ends to shore rocks, and develop their body plates.

▼ Microscopic plants (phytoplankton) are caught by the long "tentacles" of single-celled predators such as heliozoans and foraminiferans. These are the first links in the ocean's food webs, which mostly are longer and more involved than food webs on land.

Pseudopodia (flexible projections or "tentacles")

Multiple rigid shell chambers of foraminiferans

Single central chamber of star- or sun-like heliozoan

▲ Many oceanic micro-organisms have soft bodies protected inside hard, beautifully sculpted, shell-like chambers. These are constructed of natural silica minerals – the same basic substance we use to make glass.

A drip in the ocean

In fresh water, amphibian larvae, known as tadpoles, change body shape (metamorphose) to become frogs. In the sea, countless larvae do the same. But they begin smaller, take longer, and metamorphose through more, stranger stages – six or seven for some. Very few of these youngsters eventually grow up to become starfish, crabs, prawns, shrimps, sea-snails, mud-dwelling worms, jellyfish and similar marine creatures. The vast mass of the "larval soup" is destined to be food for larger, filter-feeding animals.

Body protected by jelly-like covering

Copepod swims by rapidly rowing with its long antennae

Egg sac of female copepod

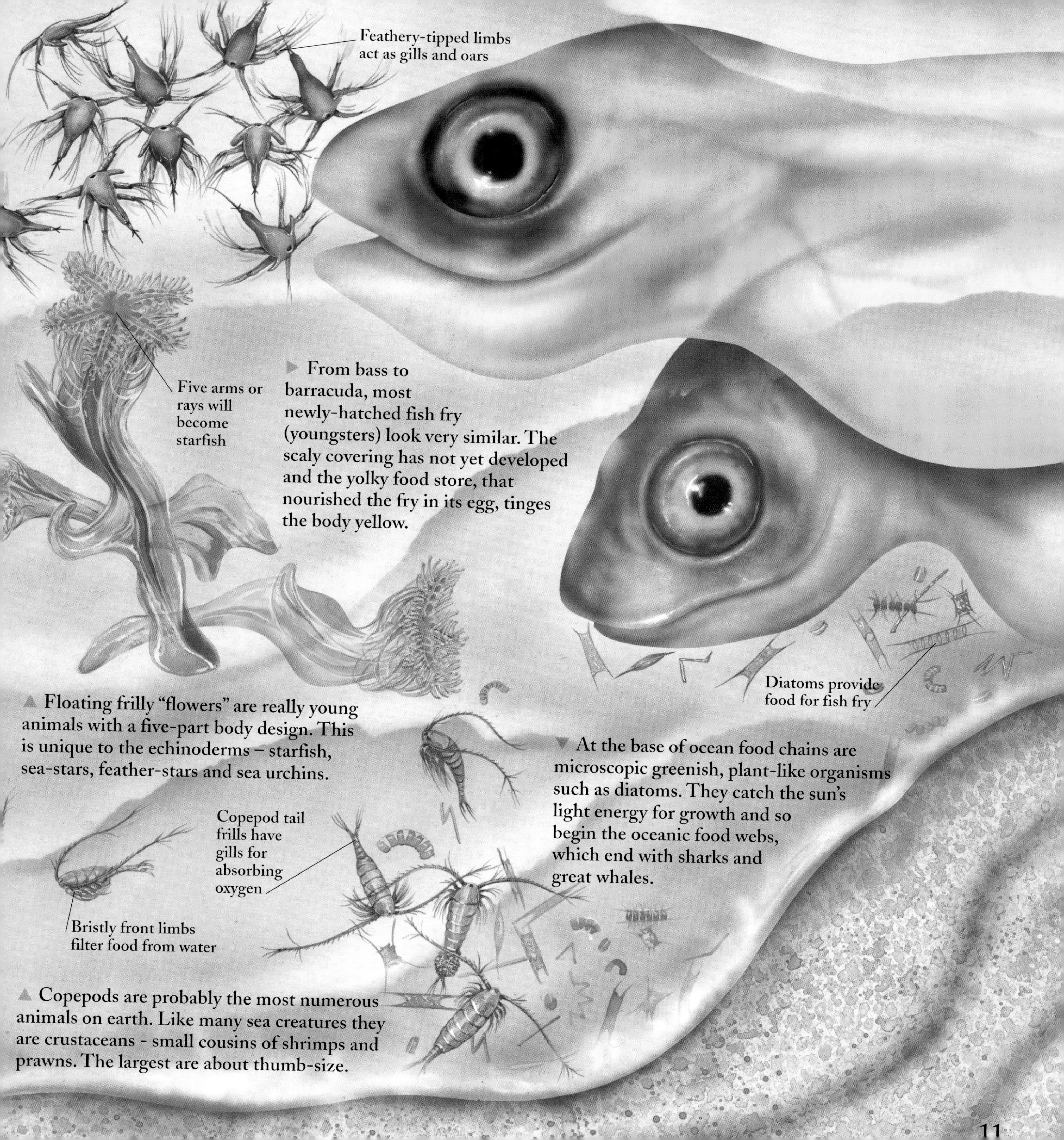
Feathery-tipped limbs act as gills and oars
Five arms or rays will become starfish
Diatoms provide food for fish fry
Copepod tail frills have gills for absorbing oxygen
Bristly front limbs filter food from water

▶ From bass to barracuda, most newly-hatched fish fry (youngsters) look very similar. The scaly covering has not yet developed and the yolky food store, that nourished the fry in its egg, tinges the body yellow.

▲ Floating frilly "flowers" are really young animals with a five-part body design. This is unique to the echinoderms – starfish, sea-stars, feather-stars and sea urchins.

▼ At the base of ocean food chains are microscopic greenish, plant-like organisms such as diatoms. They catch the sun's light energy for growth and so begin the oceanic food webs, which end with sharks and great whales.

▲ Copepods are probably the most numerous animals on earth. Like many sea creatures they are crustaceans - small cousins of shrimps and prawns. The largest are about thumb-size.

Life appears in a rain puddle within minutes. How? Falling drops wash over leaves, bark and other surfaces, and collect tiny animals, plants, eggs and spores from the thin films of moisture on them. Also, soil creatures come up from below to investigate the vast ocean forming on their ceiling.

MAGNIFICATION

x250

▶ Chlamydomonas are single-celled, plant-like organisms called flagellates. They trap sunlight and so form a basic food source.

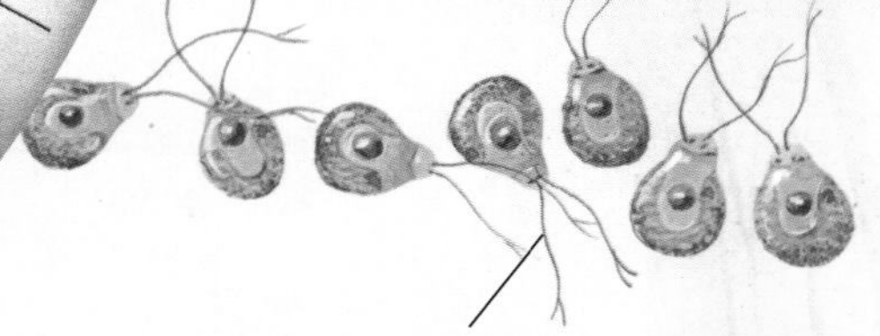

Long, whip-like flagella are waved to move to a brighter place for more light energy

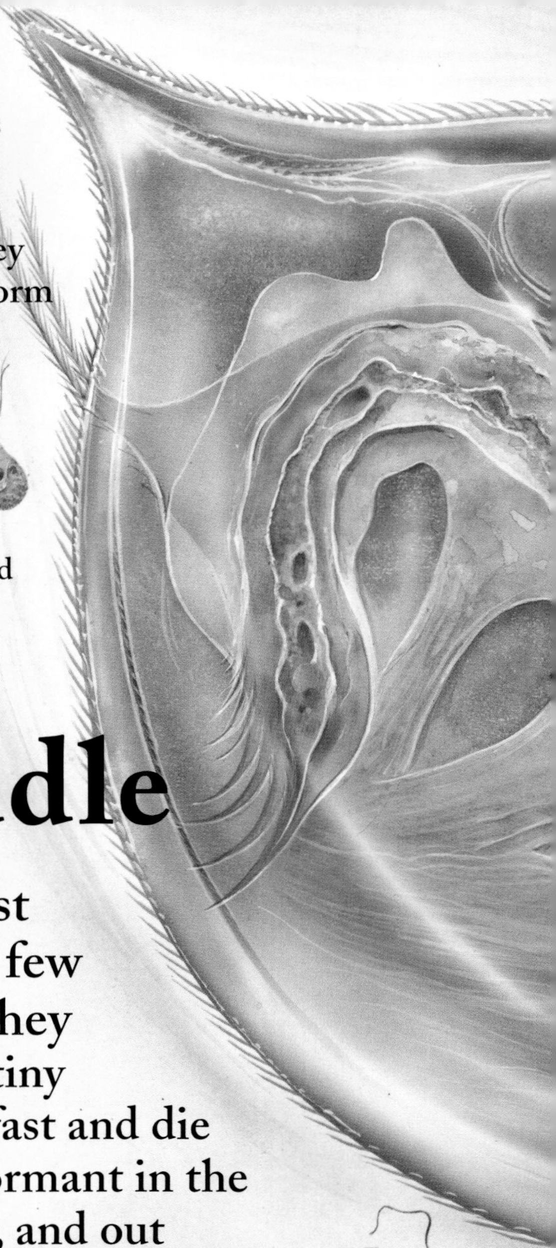

A passing puddle

Pools of rainwater last only days, even just a few hours. Yet almost as they form, they become home to tiny aquatic creatures which live fast and die young. Eggs that have lain dormant in the dust for months hatch rapidly, and out into the fresh water come mini-relatives of jellyfish, crabs and other sea animals. They begin to feed almost at once. Some eat the micro-plants that also burst into life as the rain patters down. Others eat each other.

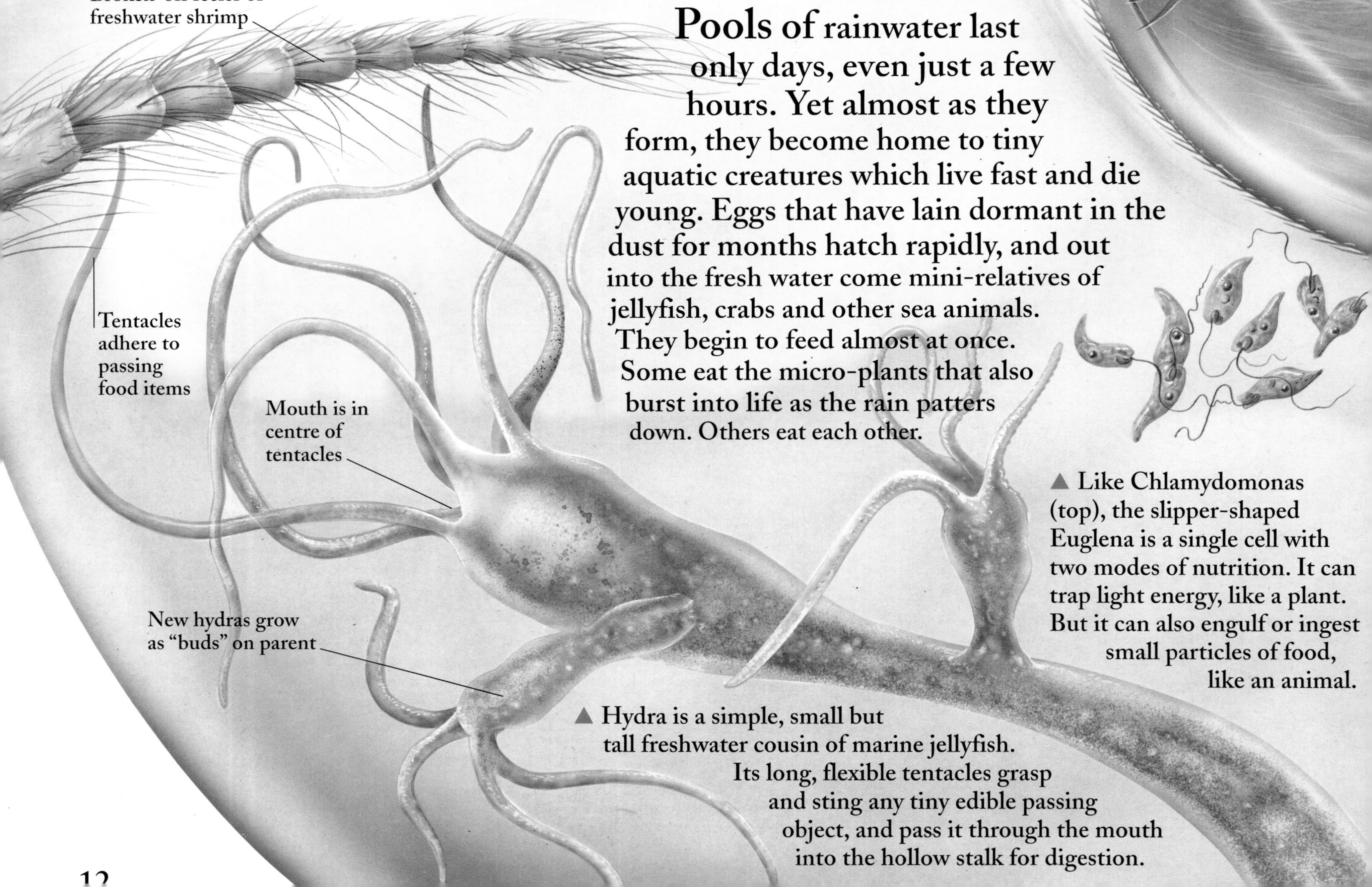

Broken-off feeler of freshwater shrimp

Tentacles adhere to passing food items

Mouth is in centre of tentacles

New hydras grow as "buds" on parent

▲ Like Chlamydomonas (top), the slipper-shaped Euglena is a single cell with two modes of nutrition. It can trap light energy, like a plant. But it can also engulf or ingest small particles of food, like an animal.

▲ Hydra is a simple, small but tall freshwater cousin of marine jellyfish. Its long, flexible tentacles grasp and sting any tiny edible passing object, and pass it through the mouth into the hollow stalk for digestion.

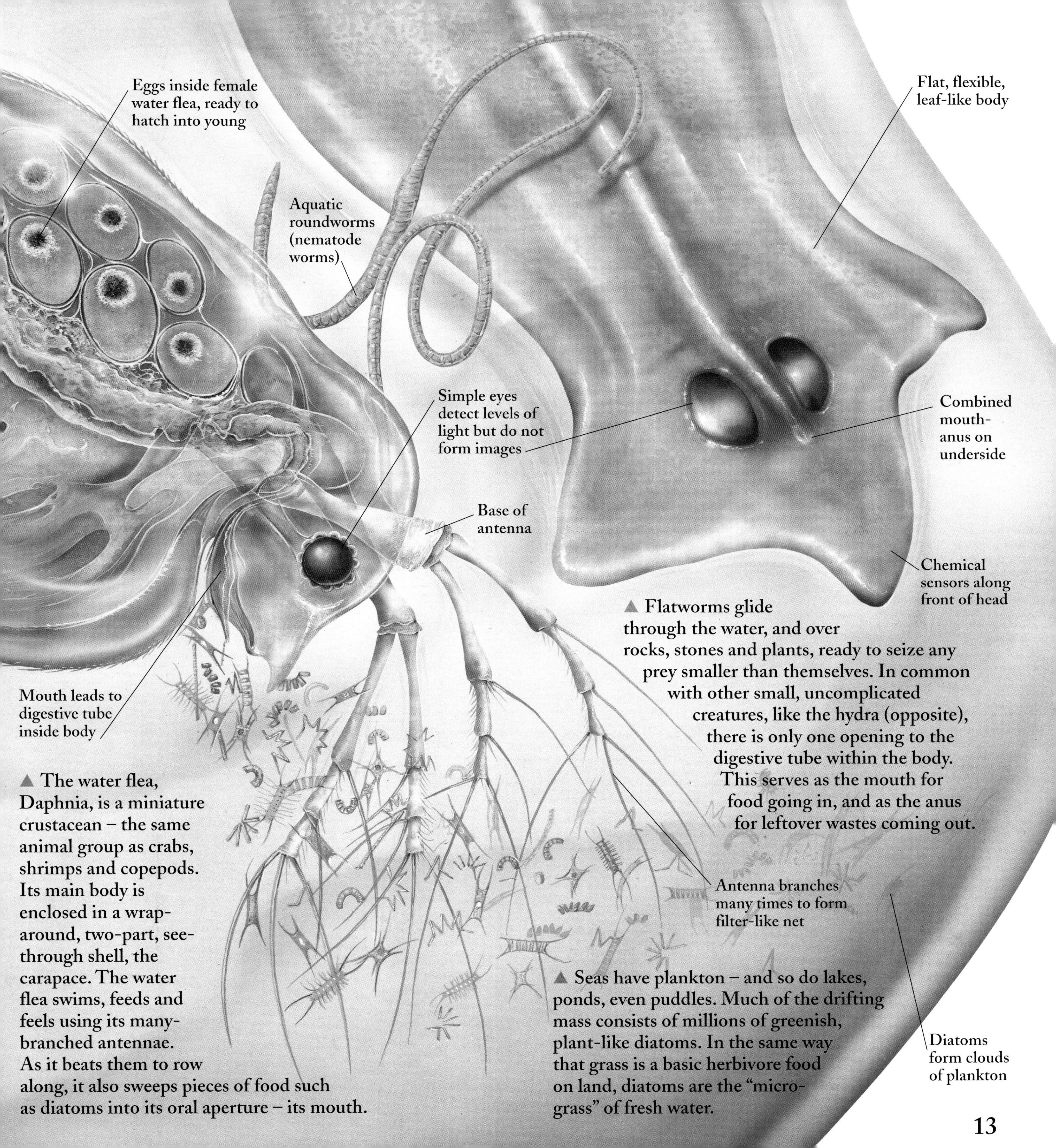

▲ Flatworms glide through the water, and over rocks, stones and plants, ready to seize any prey smaller than themselves. In common with other small, uncomplicated creatures, like the hydra (opposite), there is only one opening to the digestive tube within the body. This serves as the mouth for food going in, and as the anus for leftover wastes coming out.

▲ The water flea, Daphnia, is a miniature crustacean – the same animal group as crabs, shrimps and copepods. Its main body is enclosed in a wrap-around, two-part, see-through shell, the carapace. The water flea swims, feeds and feels using its many-branched antennae. As it beats them to row along, it also sweeps pieces of food such as diatoms into its oral aperture – its mouth.

▲ Seas have plankton – and so do lakes, ponds, even puddles. Much of the drifting mass consists of millions of greenish, plant-like diatoms. In the same way that grass is a basic herbivore food on land, diatoms are the "micro-grass" of fresh water.

Green house

A leaf seems too thin, to make a home in. Yet a whole host of creatures dwell there, spending their lives in the narrow blade, hardly thicker than this page, between the leaf's upper and lower surfaces. On the outside, the only signs of their habitation are brownish trails which mark their travels, as they chew their wandering tunnels. The leaf may react by trying to wall off its unwelcome guests, producing strange, hardened, scar-like structures variously shaped as buttons and blobs, and known as galls.

Thousands of brittle-tipped hairs coat stinging nettle
Caterpillar is thickly coated with irritant hairs
Chewing mouth-parts at head end
Body segments of caterpillar
Fluid leaking from the nettle's broken hair tip causes stinging, pain and irritation
Wasp's feet are tipped with claws for strong grip on shiny leaf surface
Waxy, protective covering (cuticle) on top and bottom of leaf
Layers of microscopic cells make up lamina (leaf blade)
Xylem tubes in vein carry water

▲ Most plants suffer from the bite, munches and chewy attentions of herbivorous animals. But stinging nettles are covered with tiny hairs which contain pain-inducing chemical weapons, so many creatures avoid them.

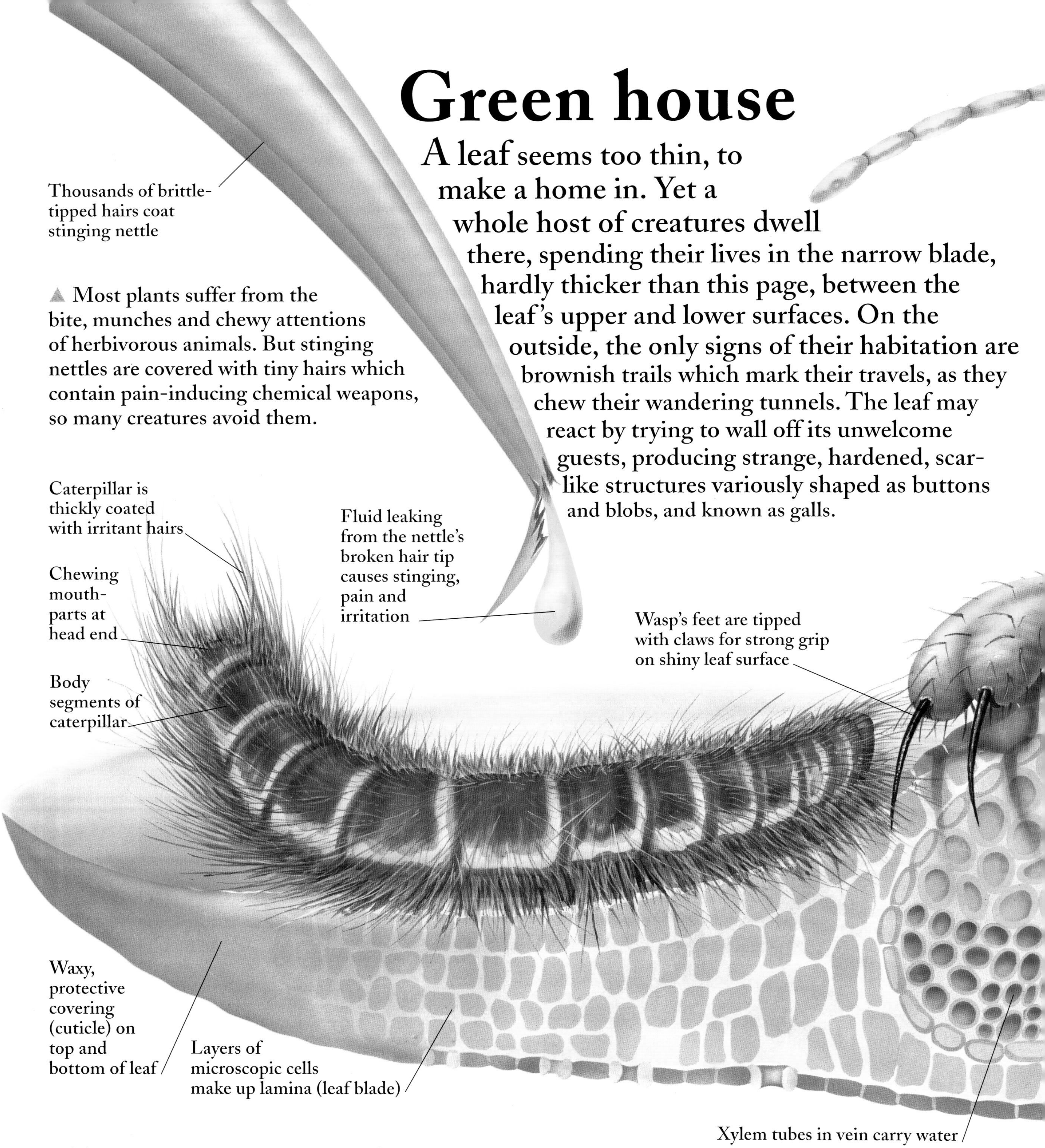

▲ The blade of a leaf has a highly ordered structure of millions of individual cells, neatly arranged in layers, with air spaces between.

▲ Some moth caterpillars are finger-sized. But thousands of kinds of small, brown moths, the geometrids, have caterpillars like pieces of dirty, hairy, wriggling cotton.

▲ The stiffer, thicker, more tubular part of the leaf is the vein. Its bundles of micro-pipes carry water and sap around the plant.

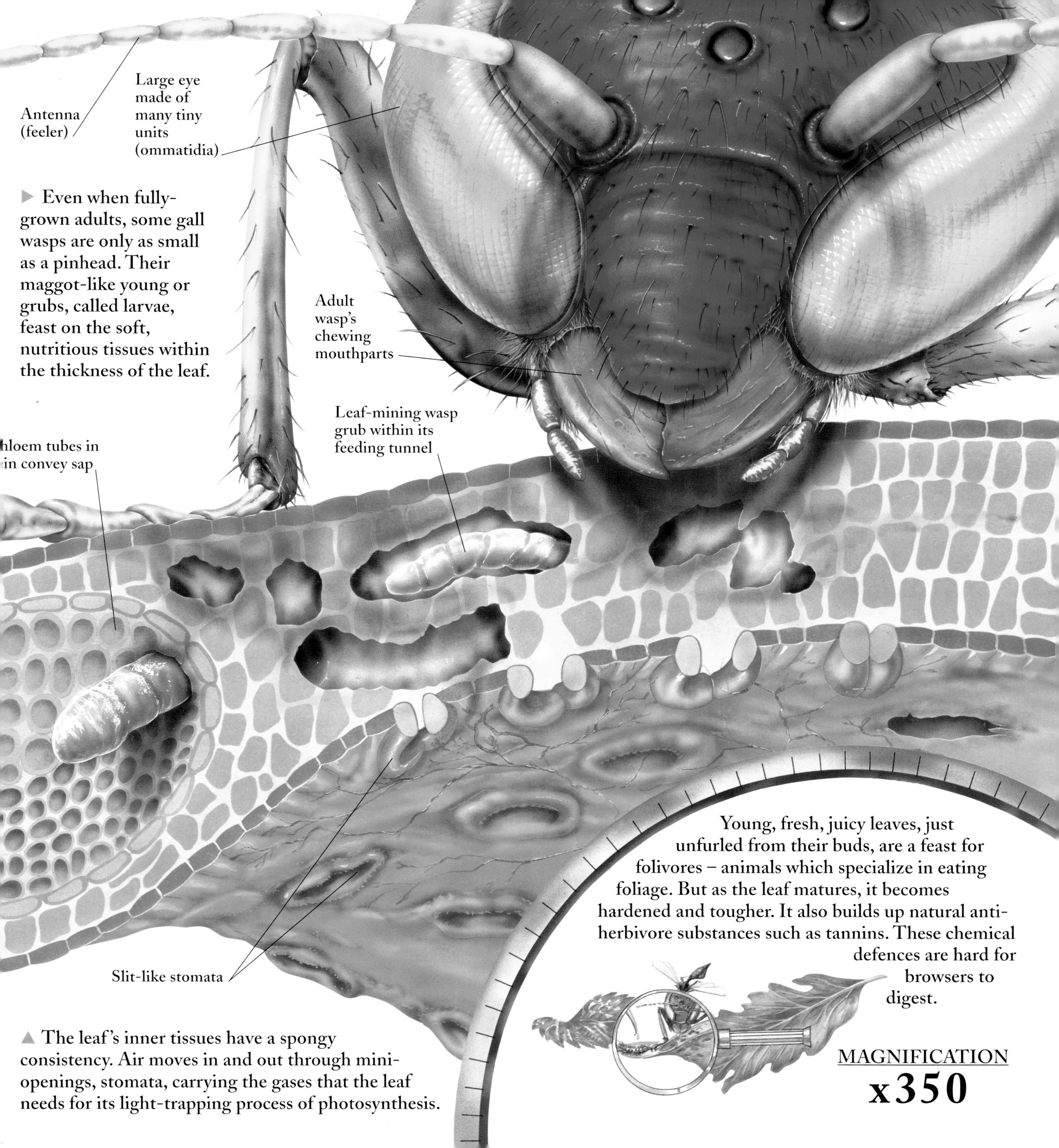

▶ Even when fully-grown adults, some gall wasps are only as small as a pinhead. Their maggot-like young or grubs, called larvae, feast on the soft, nutritious tissues within the thickness of the leaf.

Young, fresh, juicy leaves, just unfurled from their buds, are a feast for folivores – animals which specialize in eating foliage. But as the leaf matures, it becomes hardened and tougher. It also builds up natural anti-herbivore substances such as tannins. These chemical defences are hard for browsers to digest.

▲ The leaf's inner tissues have a spongy consistency. Air moves in and out through mini-openings, stomata, carrying the gases that the leaf needs for its light-trapping process of photosynthesis.

MAGNIFICATION

x350

▶ **Fruit flies turn up at almost any plant-based meal, from a newly opened bloom, to ripe and even rotting fruits. They dab up nectar, sap and similar nourishment with their sponge-like mouthparts.**

Fruit fly is a true fly, with only one pair of wings
Colourful petals attract pollinators
Tiny yellow pollen grains develop in bag-like anther (male part) of flower
Bee receives dusting of pollen at each flower visited
Pollen grains from another flower land on stigma (female part) of flower
Glossa (tongue)

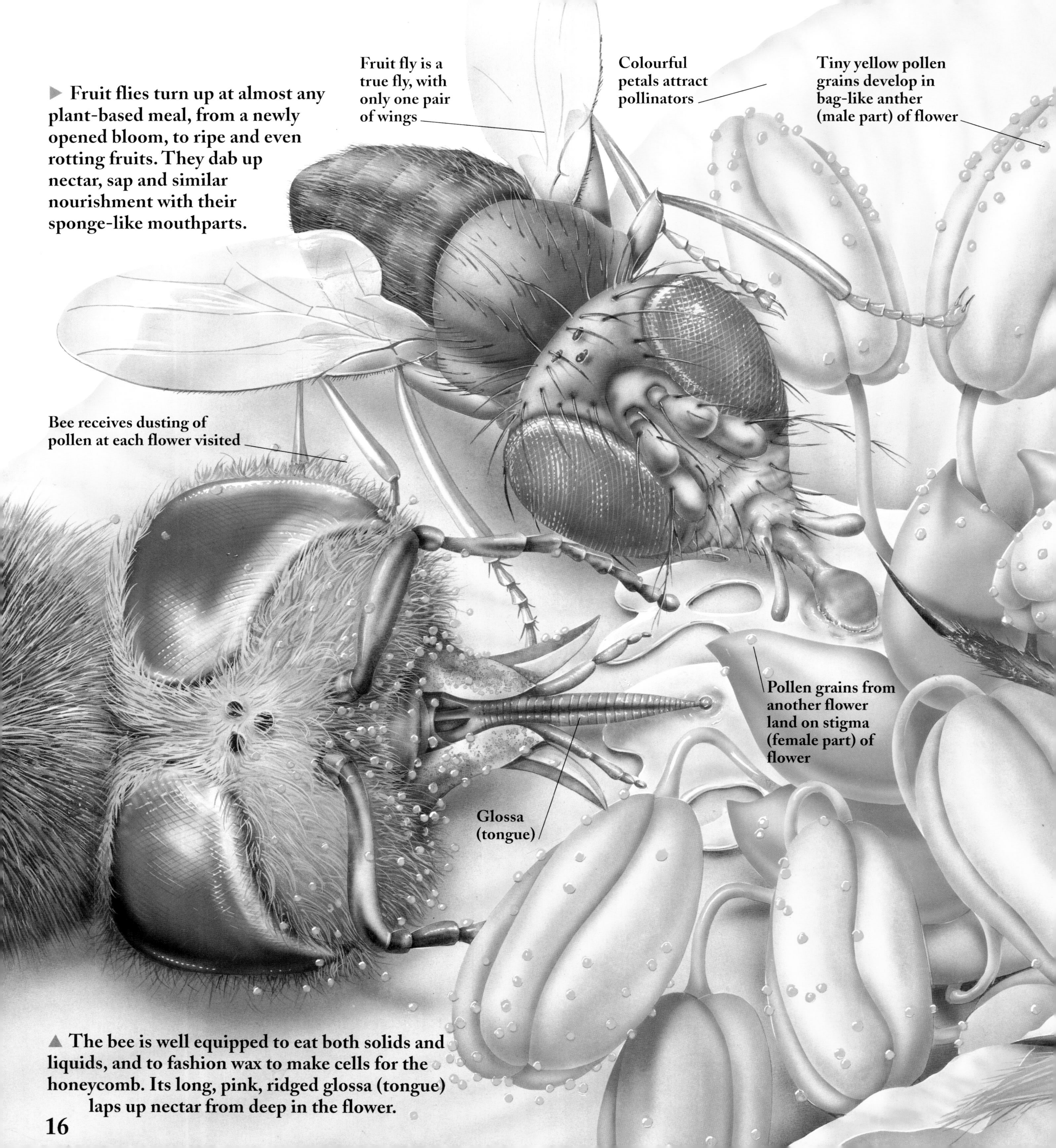

▲ **The bee is well equipped to eat both solids and liquids, and to fashion wax to make cells for the honeycomb. Its long, pink, ridged glossa (tongue) laps up nectar from deep in the flower.**

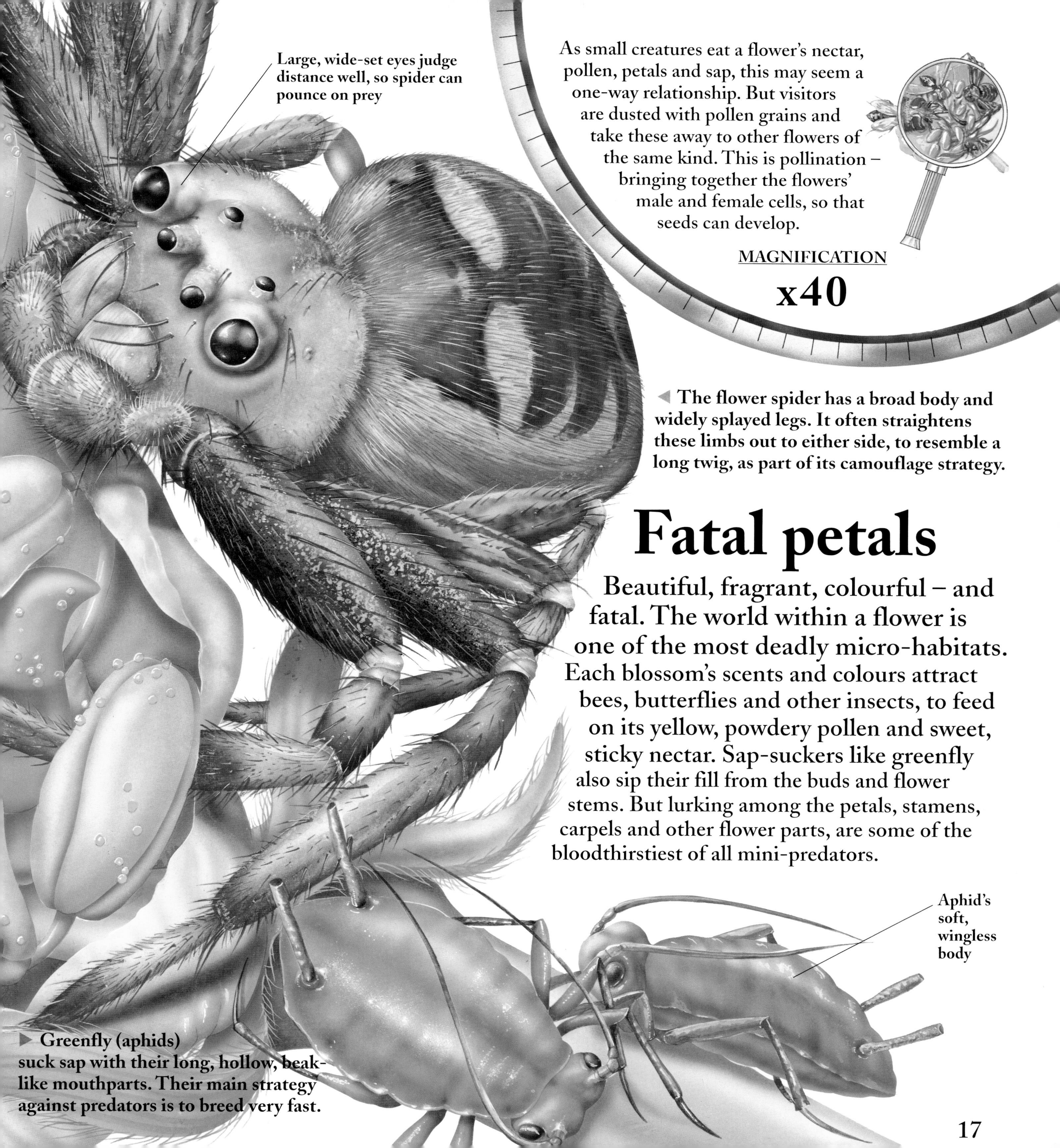

As small creatures eat a flower's nectar, pollen, petals and sap, this may seem a one-way relationship. But visitors are dusted with pollen grains and take these away to other flowers of the same kind. This is pollination – bringing together the flowers' male and female cells, so that seeds can develop.

MAGNIFICATION

x40

◀ **The flower spider has a broad body and widely splayed legs. It often straightens these limbs out to either side, to resemble a long twig, as part of its camouflage strategy.**

Fatal petals

Beautiful, fragrant, colourful – and fatal. The world within a flower is one of the most deadly micro-habitats. Each blossom's scents and colours attract bees, butterflies and other insects, to feed on its yellow, powdery pollen and sweet, sticky nectar. Sap-suckers like greenfly also sip their fill from the buds and flower stems. But lurking among the petals, stamens, carpels and other flower parts, are some of the bloodthirstiest of all mini-predators.

▶ **Greenfly (aphids) suck sap with their long, hollow, beak-like mouthparts. Their main strategy against predators is to breed very fast.**

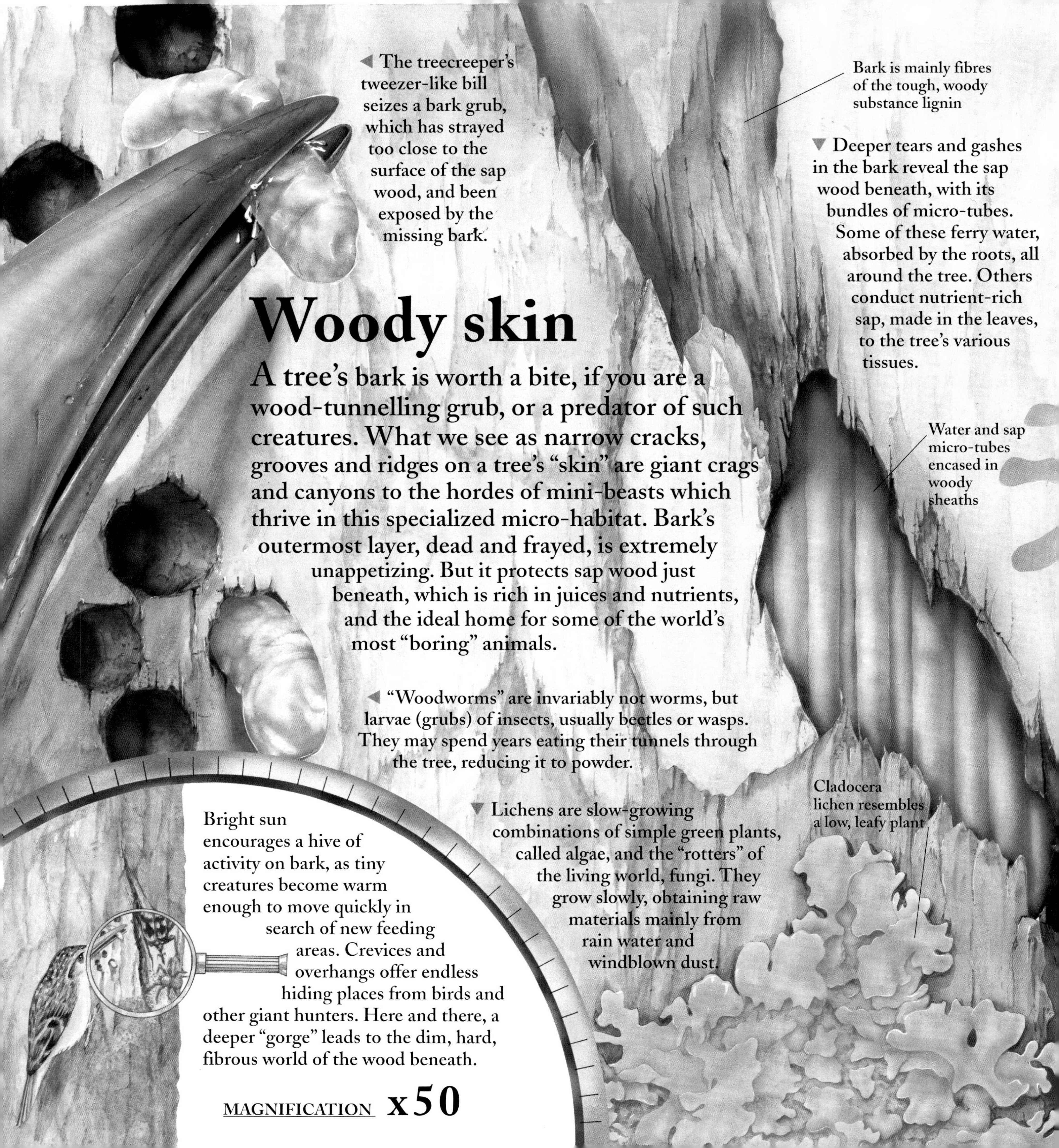
Bark is mainly fibres of the tough, woody substance lignin
Water and sap micro-tubes encased in woody sheaths
Cladocera lichen resembles a low, leafy plant

◀ The treecreeper's tweezer-like bill seizes a bark grub, which has strayed too close to the surface of the sap wood, and been exposed by the missing bark.

▼ Deeper tears and gashes in the bark reveal the sap wood beneath, with its bundles of micro-tubes. Some of these ferry water, absorbed by the roots, all around the tree. Others conduct nutrient-rich sap, made in the leaves, to the tree's various tissues.

Woody skin

A tree's bark is worth a bite, if you are a wood-tunnelling grub, or a predator of such creatures. What we see as narrow cracks, grooves and ridges on a tree's "skin" are giant crags and canyons to the hordes of mini-beasts which thrive in this specialized micro-habitat. Bark's outermost layer, dead and frayed, is extremely unappetizing. But it protects sap wood just beneath, which is rich in juices and nutrients, and the ideal home for some of the world's most "boring" animals.

◀ "Woodworms" are invariably not worms, but larvae (grubs) of insects, usually beetles or wasps. They may spend years eating their tunnels through the tree, reducing it to powder.

▼ Lichens are slow-growing combinations of simple green plants, called algae, and the "rotters" of the living world, fungi. They grow slowly, obtaining raw materials mainly from rain water and windblown dust.

Bright sun encourages a hive of activity on bark, as tiny creatures become warm enough to move quickly in search of new feeding areas. Crevices and overhangs offer endless hiding places from birds and other giant hunters. Here and there, a deeper "gorge" leads to the dim, hard, fibrous world of the wood beneath.

MAGNIFICATION **x50**

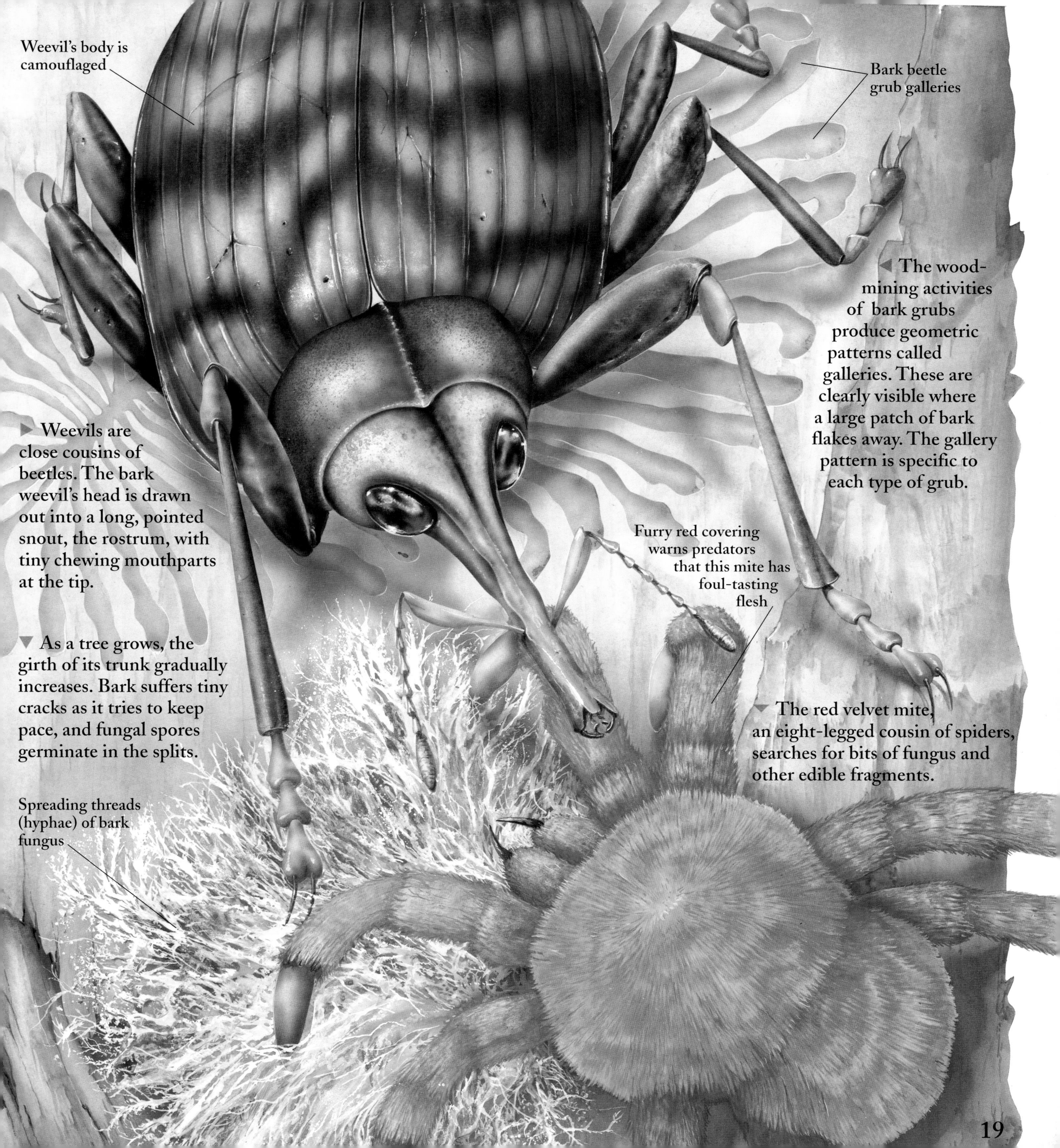

The wood-mining activities of bark grubs produce geometric patterns called galleries. These are clearly visible where a large patch of bark flakes away. The gallery pattern is specific to each type of grub.

Weevils are close cousins of beetles. The bark weevil's head is drawn out into a long, pointed snout, the rostrum, with tiny chewing mouthparts at the tip.

As a tree grows, the girth of its trunk gradually increases. Bark suffers tiny cracks as it tries to keep pace, and fungal spores germinate in the splits.

The red velvet mite, an eight-legged cousin of spiders, searches for bits of fungus and other edible fragments.

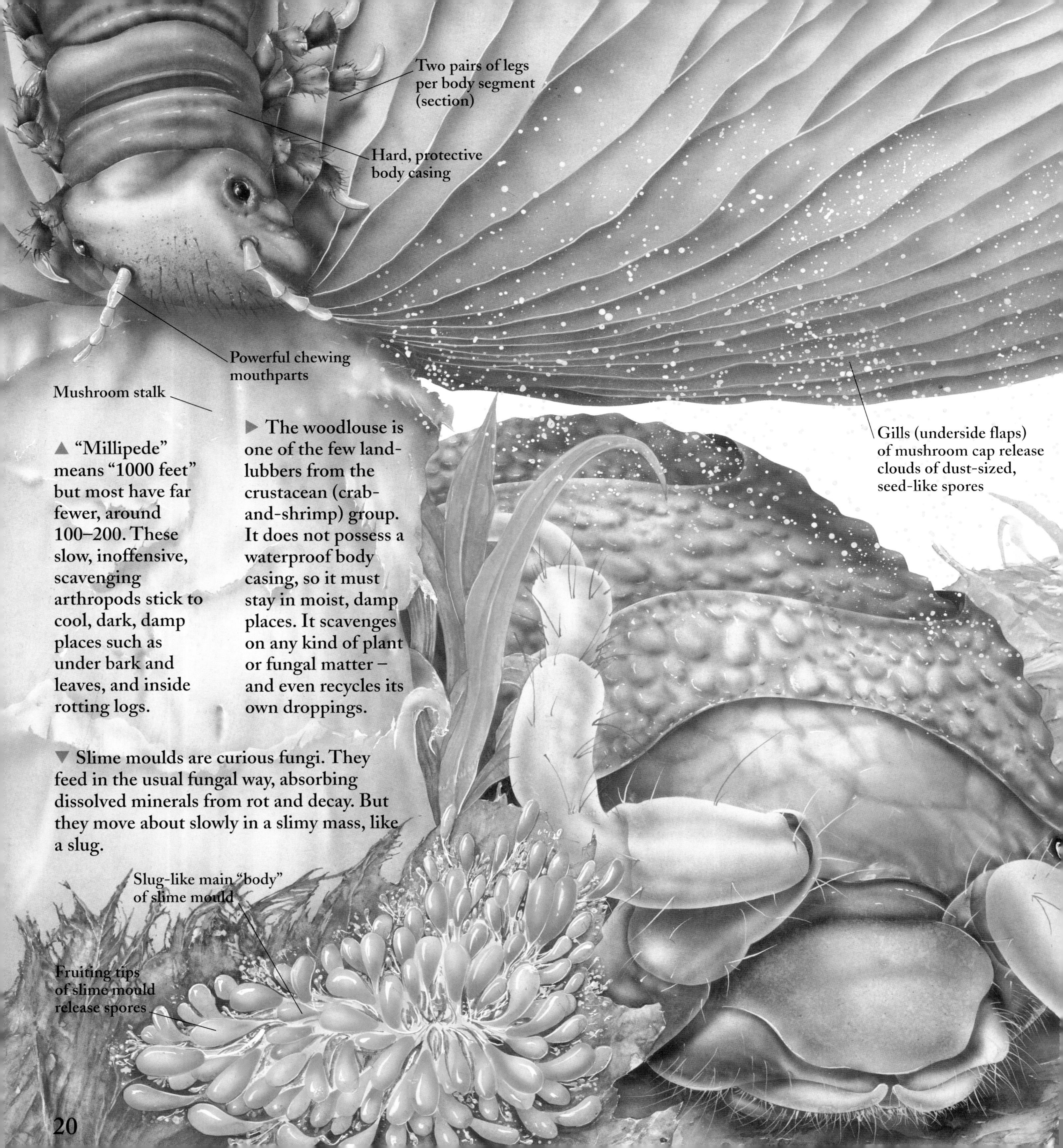

▲ "Millipede" means "1000 feet" but most have far fewer, around 100–200. These slow, inoffensive, scavenging arthropods stick to cool, dark, damp places such as under bark and leaves, and inside rotting logs.

▶ The woodlouse is one of the few land-lubbers from the crustacean (crab-and-shrimp) group. It does not possess a waterproof body casing, so it must stay in moist, damp places. It scavenges on any kind of plant or fungal matter – and even recycles its own droppings.

▼ Slime moulds are curious fungi. They feed in the usual fungal way, absorbing dissolved minerals from rot and decay. But they move about slowly in a slimy mass, like a slug.

Nature's litter

Nature sees life's struggles for survival on many levels. One of the fiercest, smallest and lowest happens in the leaf litter – the jumble of old leaves, bits of twigs, buds, seeds, nuts, fragments of flowers and fruits, animal droppings, and other discarded natural refuse beneath our feet. Here, in addition to prowling carnivores and hiding herbivores, an especially common group of creatures is the detritivores. They eat nature's leftovers (detritus) and recycle nutrients and raw materials back into the soil.

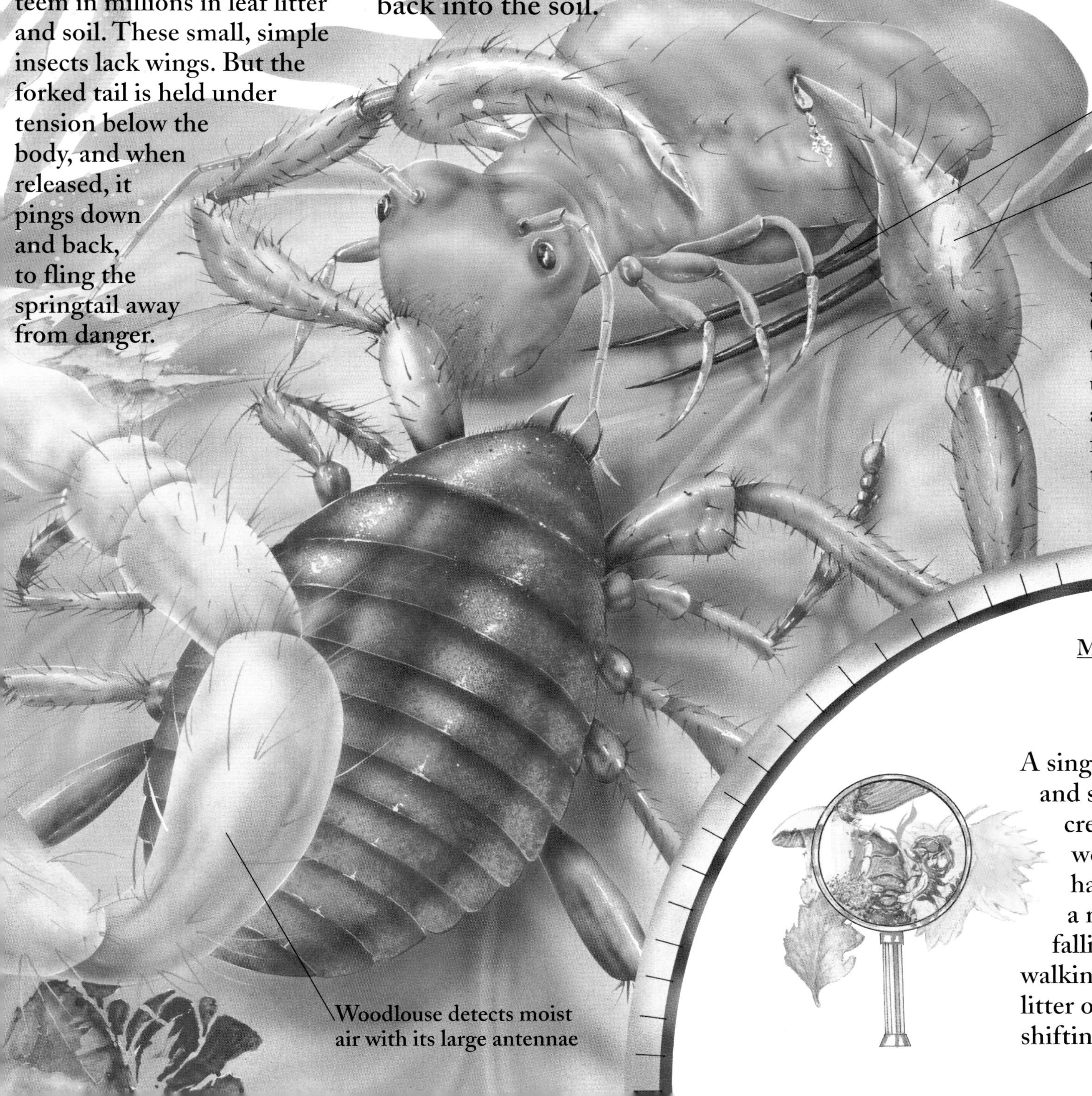

▶ Springtails (collembolans) teem in millions in leaf litter and soil. These small, simple insects lack wings. But the forked tail is held under tension below the body, and when released, it pings down and back, to fling the springtail away from danger.

Forked tail under body

False scorpion's pincers impale prey

◀ The false scorpion, or pseudoscorpion, has very scorpion-like chelicerae (pincers) – but no sting in the tail. It roams the underworld of dead leaves and nature's trash, searching for prey such as springtails, tiny grubs and worms.

Woodlouse detects moist air with its large antennae

MAGNIFICATION

x60

A single decaying leaf can feed and shelter hundreds of tiny creatures – and an average woodland tree, like an oak, has more than one-quarter of a million leaves. Wind, rain, falling twigs, and larger animals walking above, all make the leaf litter one of the most constantly shifting, changing microhabitats.

Soil

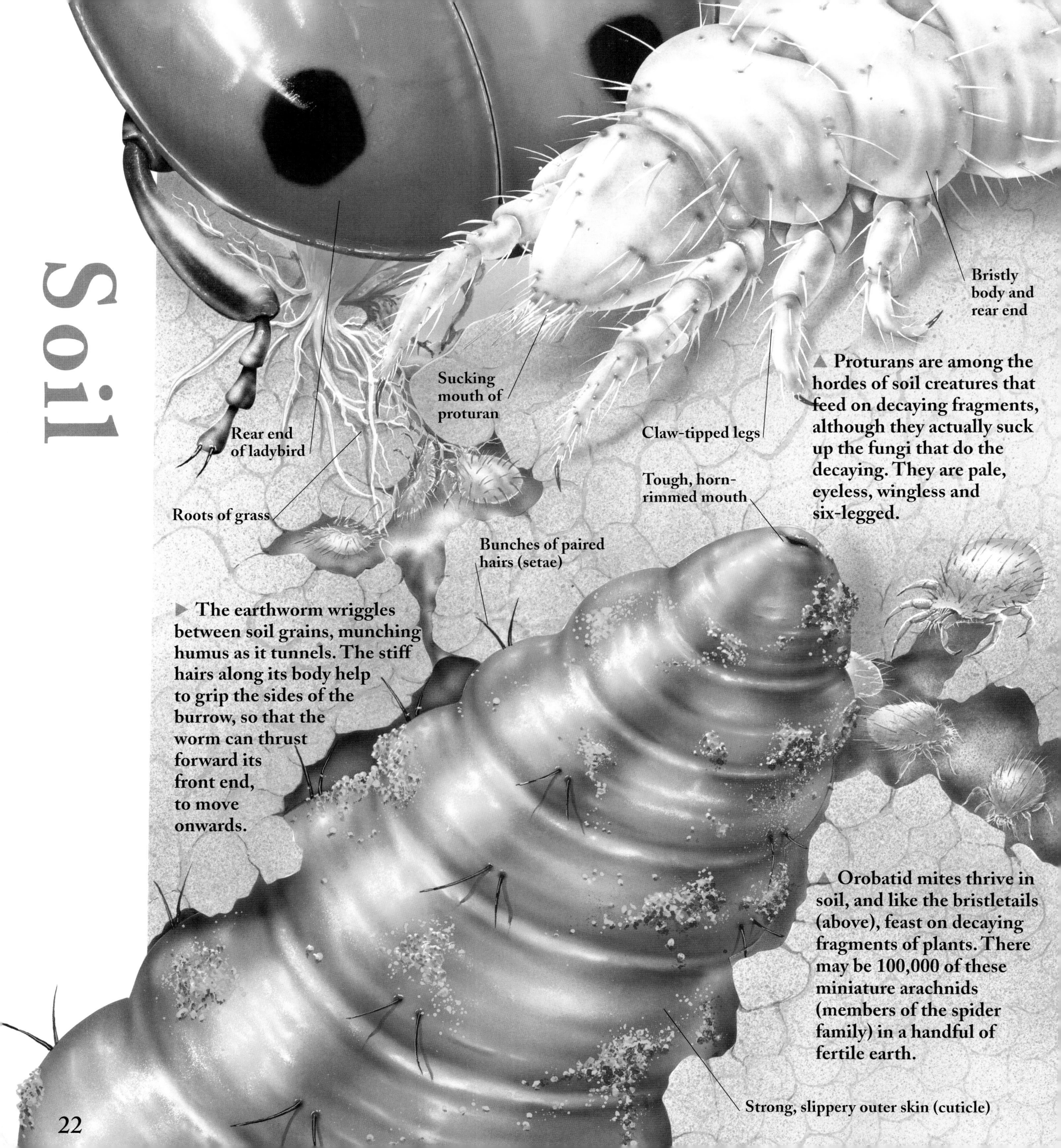

▲ Proturans are among the hordes of soil creatures that feed on decaying fragments, although they actually suck up the fungi that do the decaying. They are pale, eyeless, wingless and six-legged.

▶ The earthworm wriggles between soil grains, munching humus as it tunnels. The stiff hairs along its body help to grip the sides of the burrow, so that the worm can thrust forward its front end, to move onwards.

▲ Orobatid mites thrive in soil, and like the bristletails (above), feast on decaying fragments of plants. There may be 100,000 of these miniature arachnids (members of the spider family) in a handful of fertile earth.

MAGNIFICATION

x60

The activities of earthworms and other burrowers open up the soil's texture, allowing air and water – both vital for life – to pass into its structure. Plenty of humus and other organic matter also help, by holding water like a sponge.

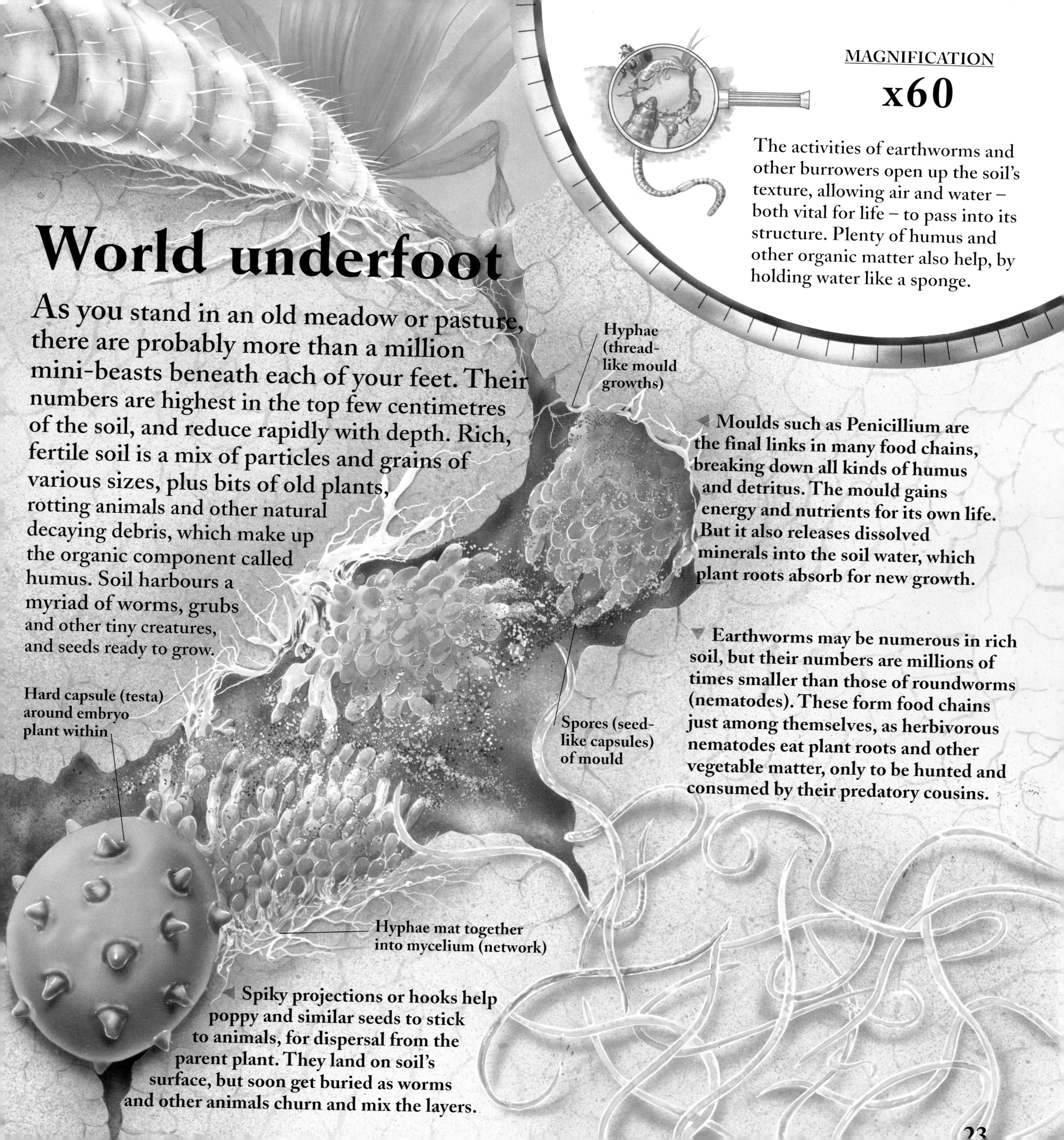

World underfoot

As you stand in an old meadow or pasture, there are probably more than a million mini-beasts beneath each of your feet. Their numbers are highest in the top few centimetres of the soil, and reduce rapidly with depth. Rich, fertile soil is a mix of particles and grains of various sizes, plus bits of old plants, rotting animals and other natural decaying debris, which make up the organic component called humus. Soil harbours a myriad of worms, grubs and other tiny creatures, and seeds ready to grow.

Hyphae (thread-like mould growths)

◀ Moulds such as Penicillium are the final links in many food chains, breaking down all kinds of humus and detritus. The mould gains energy and nutrients for its own life. But it also releases dissolved minerals into the soil water, which plant roots absorb for new growth.

▼ Earthworms may be numerous in rich soil, but their numbers are millions of times smaller than those of roundworms (nematodes). These form food chains just among themselves, as herbivorous nematodes eat plant roots and other vegetable matter, only to be hunted and consumed by their predatory cousins.

Hard capsule (testa) around embryo plant within

Spores (seed-like capsules) of mould

Hyphae mat together into mycelium (network)

◀ Spiky projections or hooks help poppy and similar seeds to stick to animals, for dispersal from the parent plant. They land on soil's surface, but soon get buried as worms and other animals churn and mix the layers.

▶ King of the predators in the plant root micro-habitat is the burrowing centipede. It wriggles and "swims" between soil particles, pushing with its legs through tighter spots. To lessen effort, it often follows the ready-made tunnel of an earthworm.

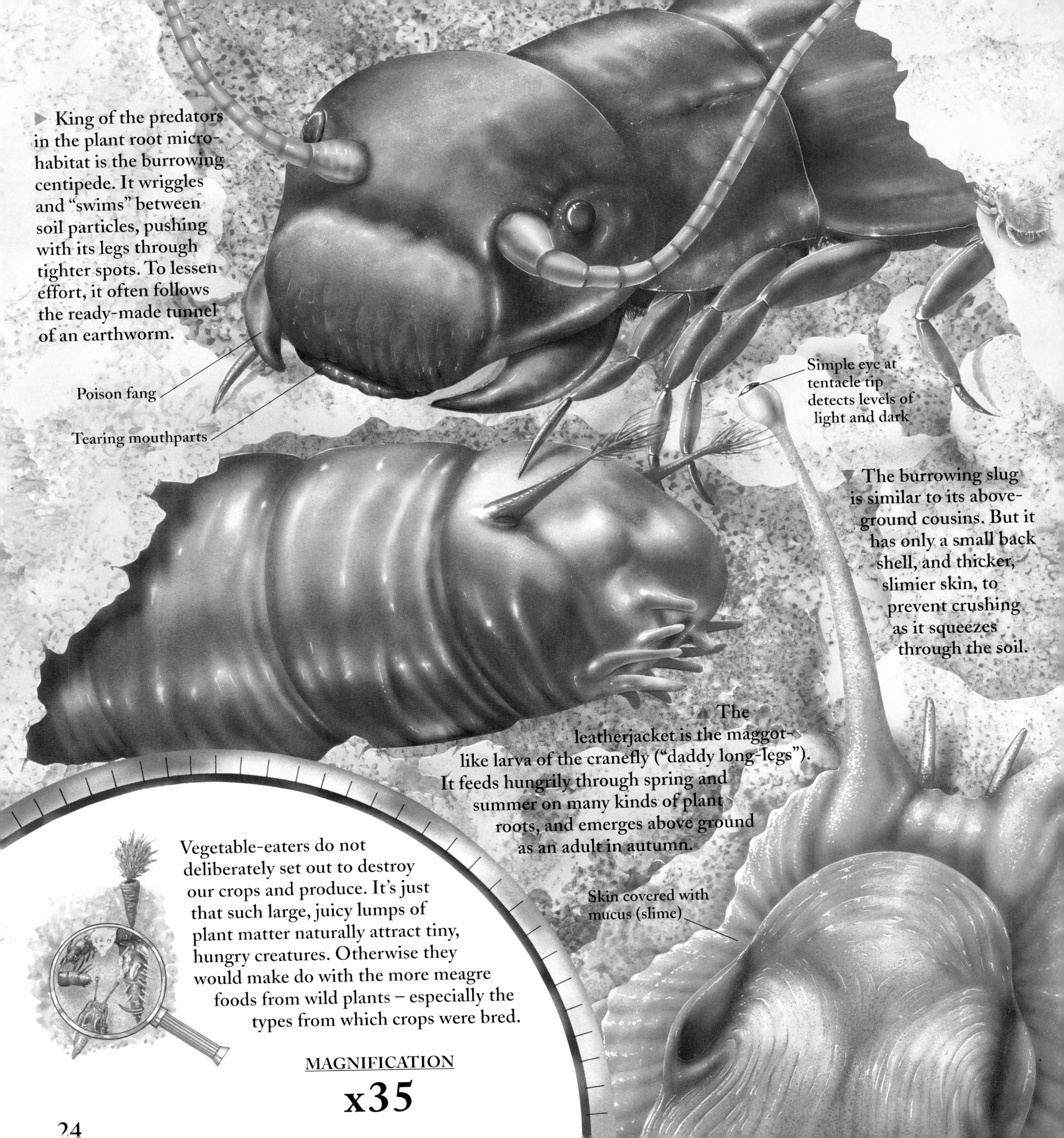

Poison fang

Tearing mouthparts

Simple eye at tentacle tip detects levels of light and dark

▼ The burrowing slug is similar to its above-ground cousins. But it has only a small back shell, and thicker, slimier skin, to prevent crushing as it squeezes through the soil.

▲ The leatherjacket is the maggot-like larva of the cranefly ("daddy long-legs"). It feeds hungrily through spring and summer on many kinds of plant roots, and emerges above ground as an adult in autumn.

Skin covered with mucus (slime)

Vegetable-eaters do not deliberately set out to destroy our crops and produce. It's just that such large, juicy lumps of plant matter naturally attract tiny, hungry creatures. Otherwise they would make do with the more meagre foods from wild plants – especially the types from which crops were bred.

MAGNIFICATION

x35

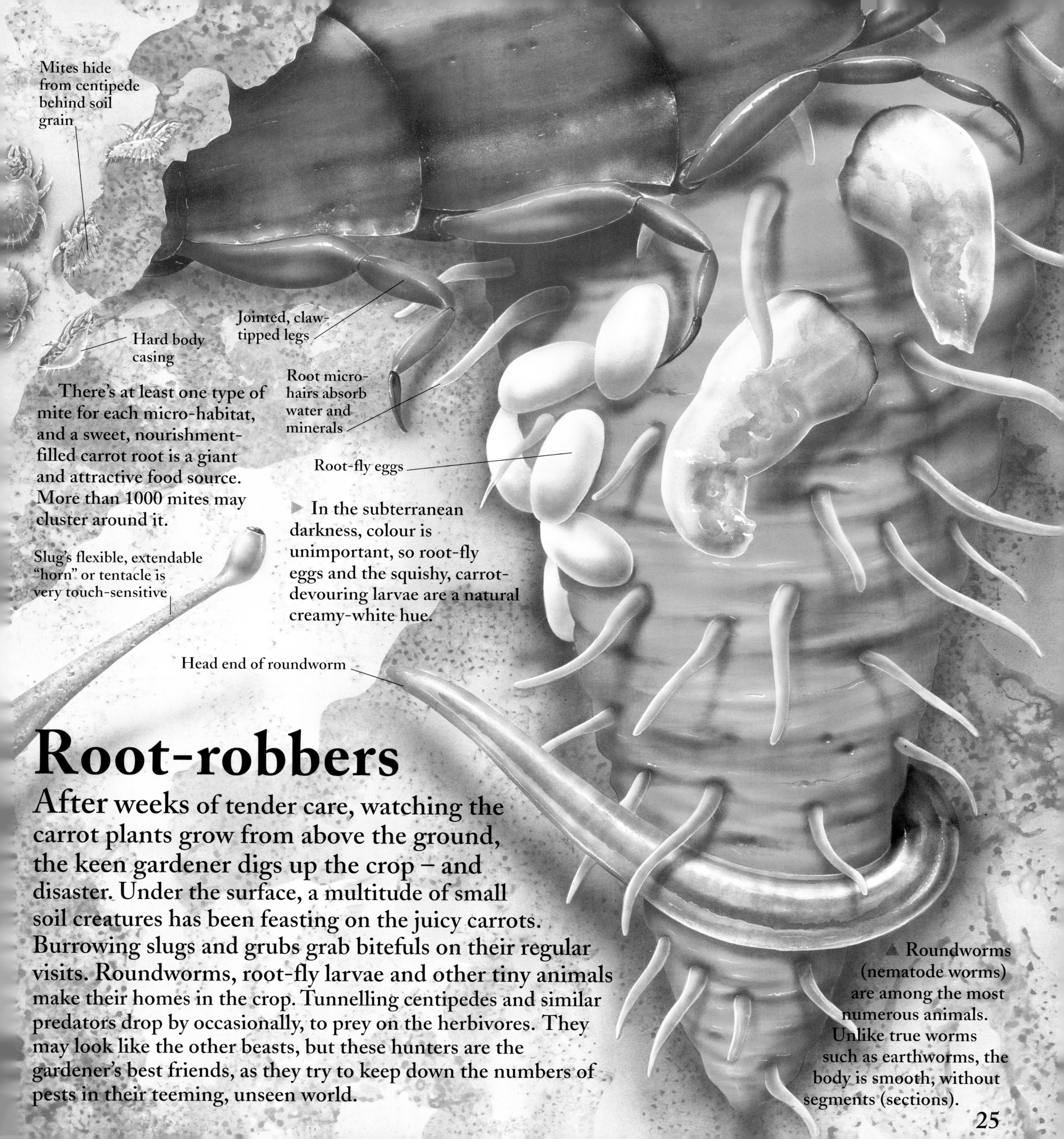

Mites hide from centipede behind soil grain

Hard body casing

Jointed, claw-tipped legs

Root micro-hairs absorb water and minerals

▲ There's at least one type of mite for each micro-habitat, and a sweet, nourishment-filled carrot root is a giant and attractive food source. More than 1000 mites may cluster around it.

Root-fly eggs

▶ In the subterranean darkness, colour is unimportant, so root-fly eggs and the squishy, carrot-devouring larvae are a natural creamy-white hue.

Slug's flexible, extendable "horn" or tentacle is very touch-sensitive

Head end of roundworm

Root-robbers

After weeks of tender care, watching the carrot plants grow from above the ground, the keen gardener digs up the crop – and disaster. Under the surface, a multitude of small soil creatures has been feasting on the juicy carrots. Burrowing slugs and grubs grab bitefuls on their regular visits. Roundworms, root-fly larvae and other tiny animals make their homes in the crop. Tunnelling centipedes and similar predators drop by occasionally, to prey on the herbivores. They may look like the other beasts, but these hunters are the gardener's best friends, as they try to keep down the numbers of pests in their teeming, unseen world.

▲ Roundworms (nematode worms) are among the most numerous animals. Unlike true worms such as earthworms, the body is smooth, without segments (sections).

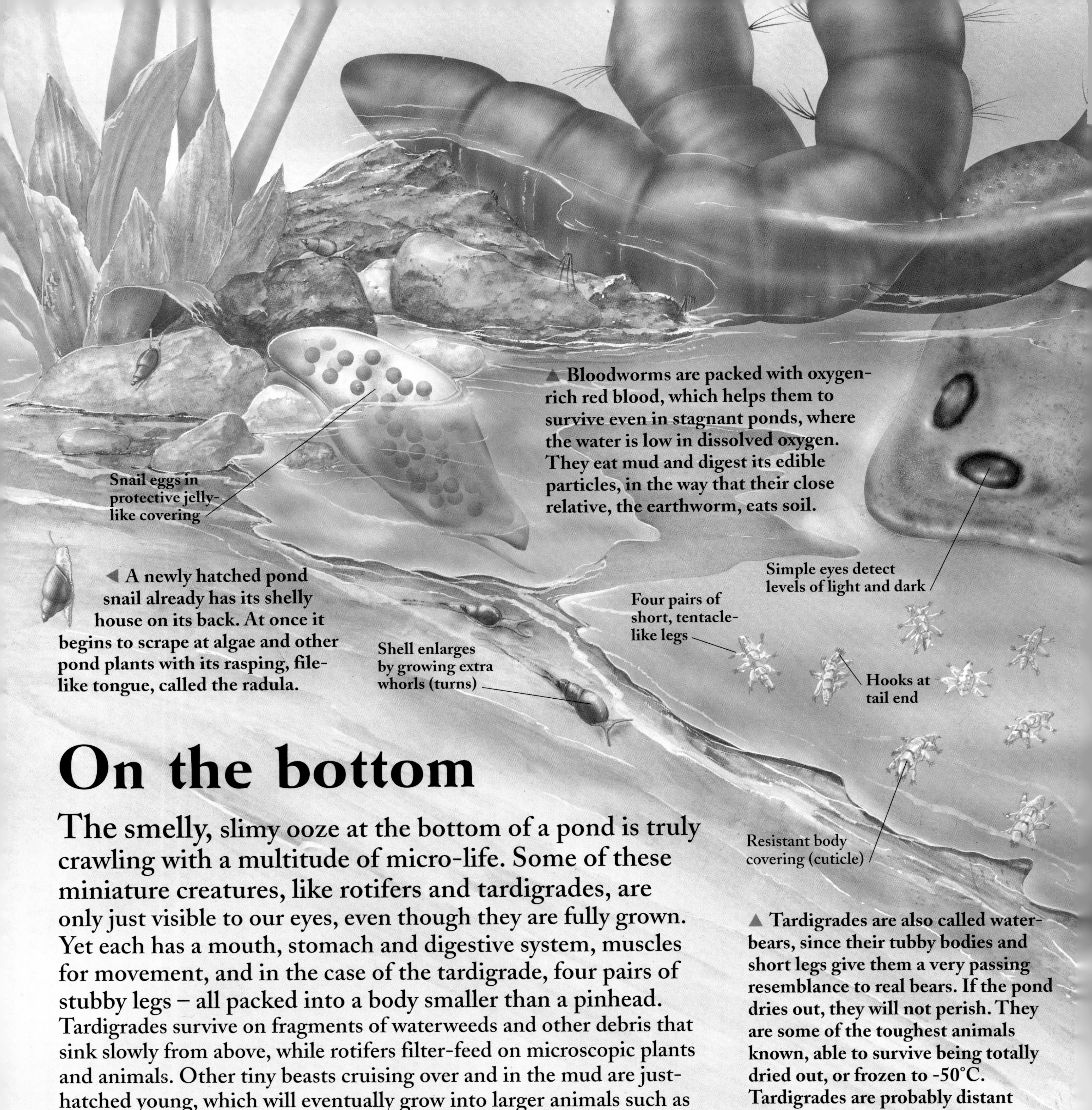

▲ Bloodworms are packed with oxygen-rich red blood, which helps them to survive even in stagnant ponds, where the water is low in dissolved oxygen. They eat mud and digest its edible particles, in the way that their close relative, the earthworm, eats soil.

◀ A newly hatched pond snail already has its shelly house on its back. At once it begins to scrape at algae and other pond plants with its rasping, file-like tongue, called the radula.

On the bottom

The smelly, slimy ooze at the bottom of a pond is truly crawling with a multitude of micro-life. Some of these miniature creatures, like rotifers and tardigrades, are only just visible to our eyes, even though they are fully grown. Yet each has a mouth, stomach and digestive system, muscles for movement, and in the case of the tardigrade, four pairs of stubby legs – all packed into a body smaller than a pinhead. Tardigrades survive on fragments of waterweeds and other debris that sink slowly from above, while rotifers filter-feed on microscopic plants and animals. Other tiny beasts cruising over and in the mud are just-hatched young, which will eventually grow into larger animals such as pond snails and damselflies – if they survive.

▲ Tardigrades are also called water-bears, since their tubby bodies and short legs give them a very passing resemblance to real bears. If the pond dries out, they will not perish. They are some of the toughest animals known, able to survive being totally dried out, or frozen to -50°C. Tardigrades are probably distant cousins of insects and spiders.

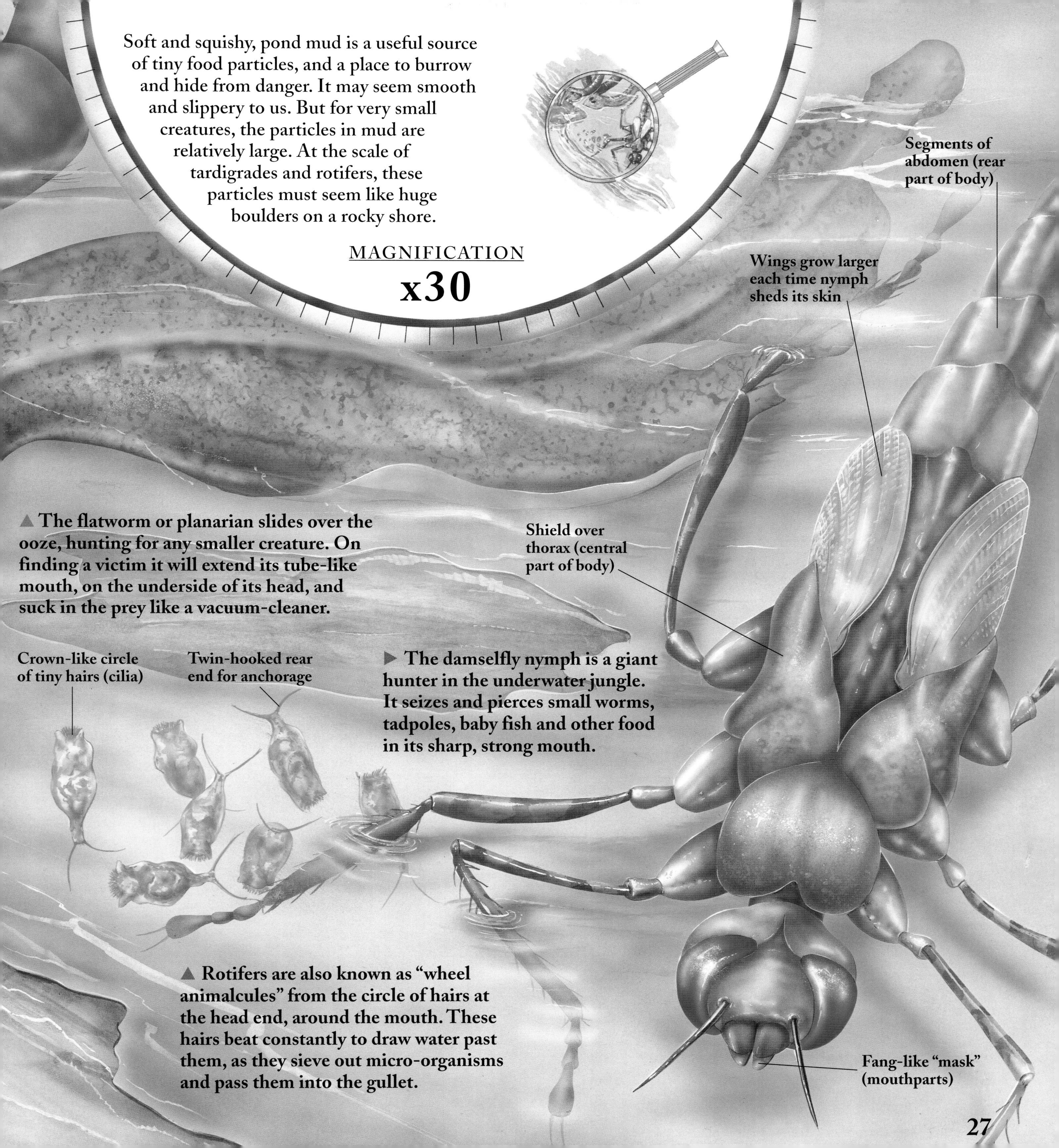

Soft and squishy, pond mud is a useful source of tiny food particles, and a place to burrow and hide from danger. It may seem smooth and slippery to us. But for very small creatures, the particles in mud are relatively large. At the scale of tardigrades and rotifers, these particles must seem like huge boulders on a rocky shore.

MAGNIFICATION

x30

▲ The flatworm or planarian slides over the ooze, hunting for any smaller creature. On finding a victim it will extend its tube-like mouth, on the underside of its head, and suck in the prey like a vacuum-cleaner.

▶ The damselfly nymph is a giant hunter in the underwater jungle. It seizes and pierces small worms, tadpoles, baby fish and other food in its sharp, strong mouth.

▲ Rotifers are also known as "wheel animalcules" from the circle of hairs at the head end, around the mouth. These hairs beat constantly to draw water past them, as they sieve out micro-organisms and pass them into the gullet.

House and Human

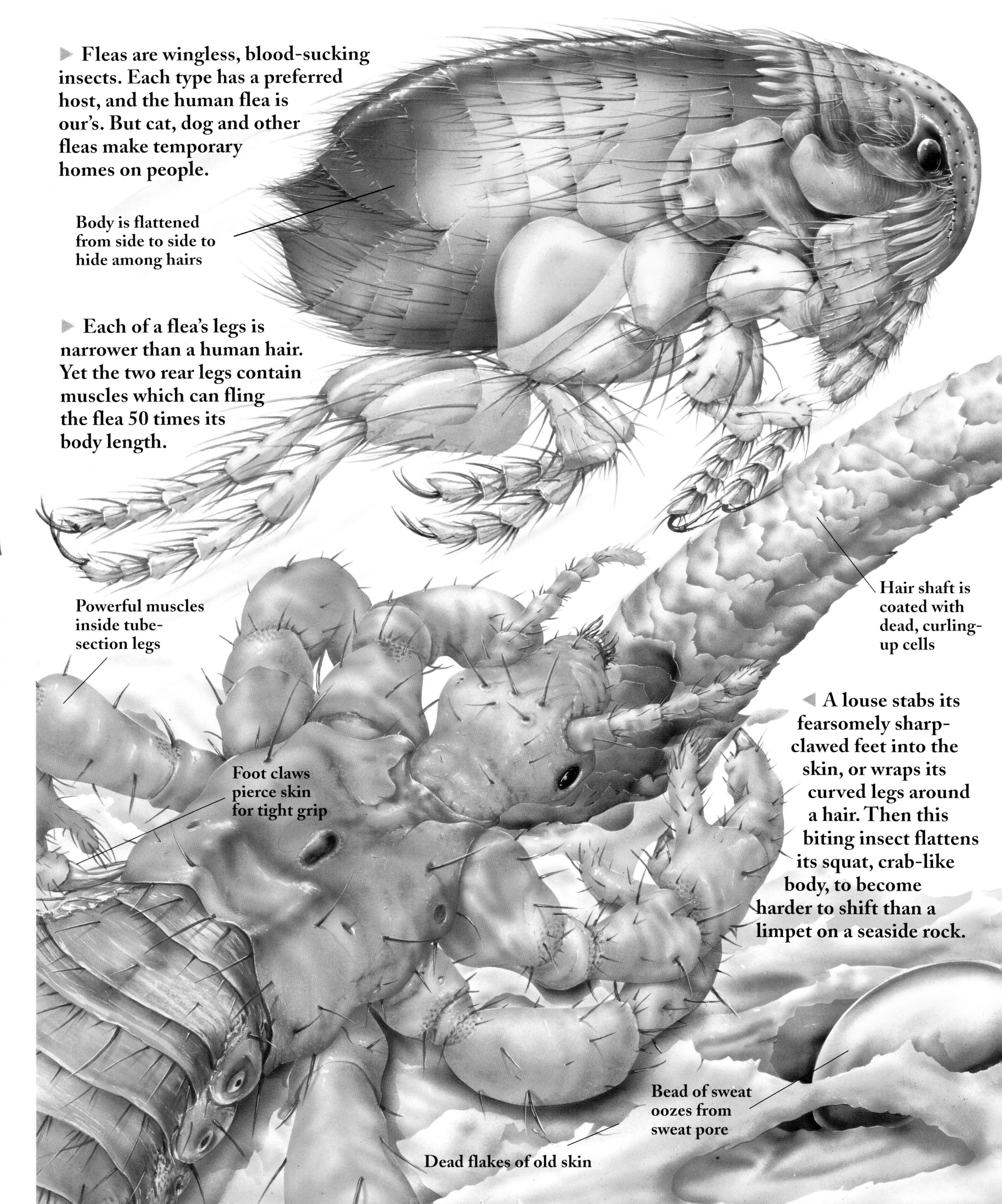

▶ Fleas are wingless, blood-sucking insects. Each type has a preferred host, and the human flea is our's. But cat, dog and other fleas make temporary homes on people.

▶ Each of a flea's legs is narrower than a human hair. Yet the two rear legs contain muscles which can fling the flea 50 times its body length.

◀ A louse stabs its fearsomely sharp-clawed feet into the skin, or wraps its curved legs around a hair. Then this biting insect flattens its squat, crab-like body, to become harder to shift than a limpet on a seaside rock.

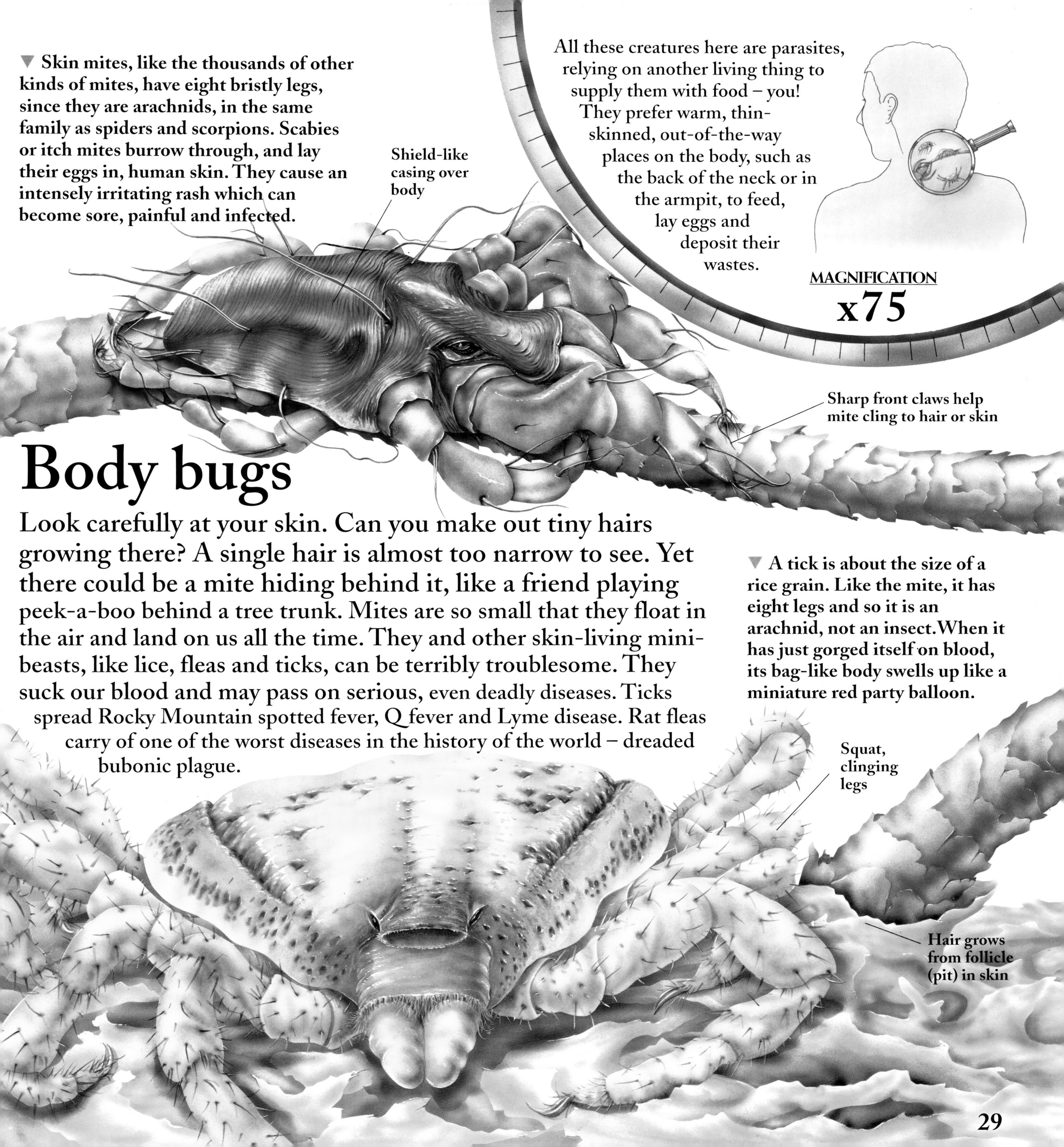

Body bugs

Look carefully at your skin. Can you make out tiny hairs growing there? A single hair is almost too narrow to see. Yet there could be a mite hiding behind it, like a friend playing peek-a-boo behind a tree trunk. Mites are so small that they float in the air and land on us all the time. They and other skin-living mini-beasts, like lice, fleas and ticks, can be terribly troublesome. They suck our blood and may pass on serious, even deadly diseases. Ticks spread Rocky Mountain spotted fever, Q fever and Lyme disease. Rat fleas carry of one of the worst diseases in the history of the world – dreaded bubonic plague.

▼ Skin mites, like the thousands of other kinds of mites, have eight bristly legs, since they are arachnids, in the same family as spiders and scorpions. Scabies or itch mites burrow through, and lay their eggs in, human skin. They cause an intensely irritating rash which can become sore, painful and infected.

All these creatures here are parasites, relying on another living thing to supply them with food – you! They prefer warm, thin-skinned, out-of-the-way places on the body, such as the back of the neck or in the armpit, to feed, lay eggs and deposit their wastes.

▼ A tick is about the size of a rice grain. Like the mite, it has eight legs and so it is an arachnid, not an insect. When it has just gorged itself on blood, its bag-like body swells up like a miniature red party balloon.

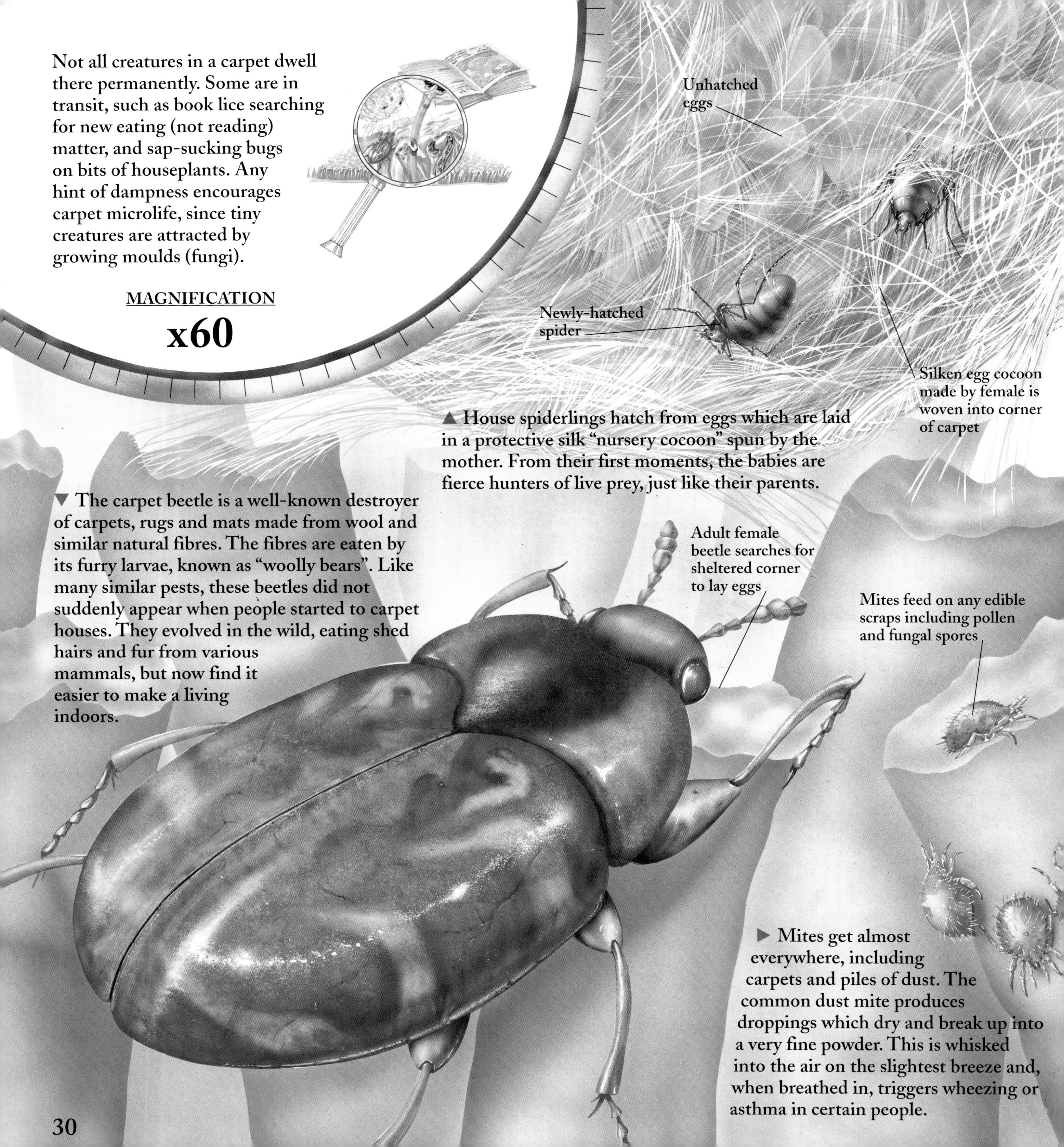
MAGNIFICATION
x60
Unhatched eggs
Newly-hatched spider
Silken egg cocoon made by female is woven into corner of carpet
Adult female beetle searches for sheltered corner to lay eggs
Mites feed on any edible scraps including pollen and fungal spores

Not all creatures in a carpet dwell there permanently. Some are in transit, such as book lice searching for new eating (not reading) matter, and sap-sucking bugs on bits of houseplants. Any hint of dampness encourages carpet microlife, since tiny creatures are attracted by growing moulds (fungi).

▲ House spiderlings hatch from eggs which are laid in a protective silk "nursery cocoon" spun by the mother. From their first moments, the babies are fierce hunters of live prey, just like their parents.

▼ The carpet beetle is a well-known destroyer of carpets, rugs and mats made from wool and similar natural fibres. The fibres are eaten by its furry larvae, known as "woolly bears". Like many similar pests, these beetles did not suddenly appear when people started to carpet houses. They evolved in the wild, eating shed hairs and fur from various mammals, but now find it easier to make a living indoors.

▶ Mites get almost everywhere, including carpets and piles of dust. The common dust mite produces droppings which dry and break up into a very fine powder. This is whisked into the air on the slightest breeze and, when breathed in, triggers wheezing or asthma in certain people.

House guests

Every time you walk across a carpet, you probably tread on hundreds of minibeasts. Even the most powerful vacuum-cleaner has trouble sucking them up. They hide among the hairs and fibres of the carpet's pile, clinging firmly, like larger creatures sheltering from a gale in a thick forest. But what do they eat, in this tangled world? Quite often – bits of you. Flakes of skin and loose hairs, also pieces of pet fur, food crumbs, bits of soil off shoes, general specks of dust … all of these tiny items contain scraps of nourishment. And if a carpet has natural fibres like wool, rather than artificial ones such as nylon, the minibeasts will eat those, too!

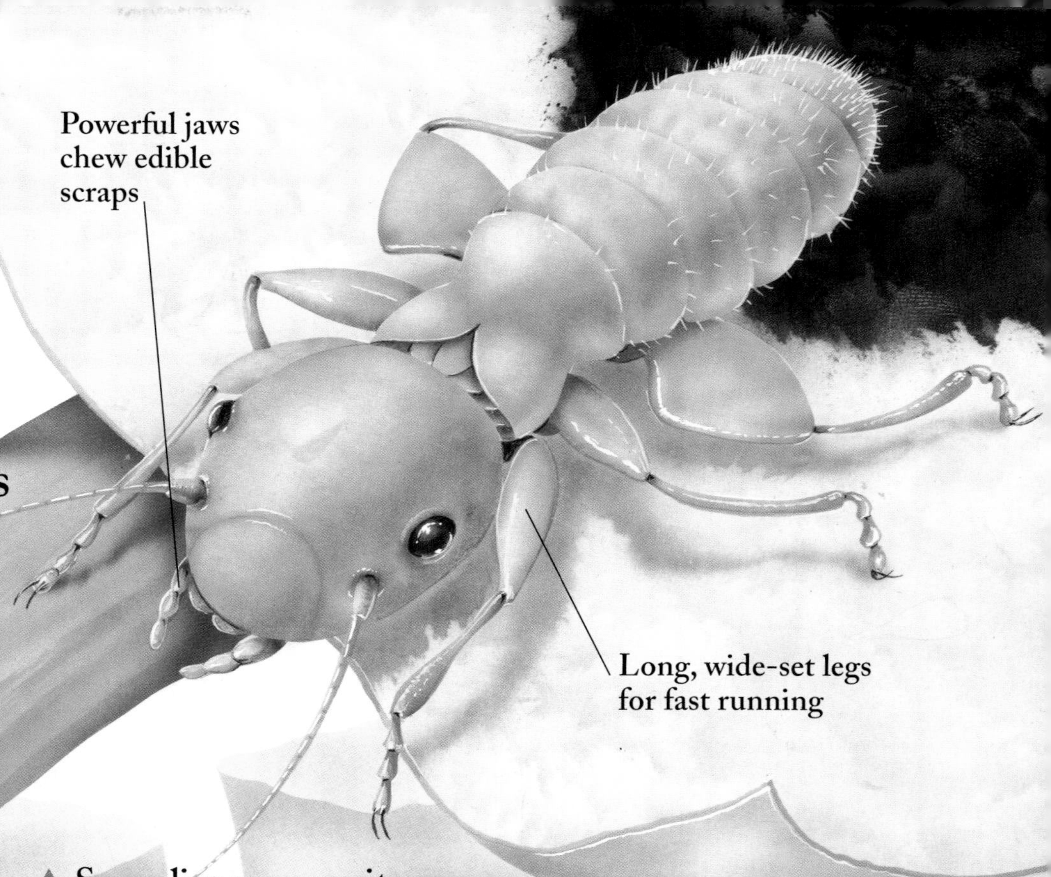

Powerful jaws chew edible scraps

Long, wide-set legs for fast running

▲ Some lice are parasites, sucking blood and fluids from larger hosts. But book lice are "free-living" (not parasitic). They were named because they fed on moulds (fungi) growing on the paper of old books, and on the glues used to bind the pages, the glues being made from animal carcasses.

Protective shell-like scale is fixed in place so adult female cannot move

Body of female scale insect

◀ Scale insects belong to the bug insect group (being cousins of greenfly). In their younger stages, they move about like normal insects. But the adult female grows a shell-like body covering that fuses to a plant stem, and feeds by sucking sap from the stem through her needle-like mouthparts.

Strong jaw-like mandibles for grinding and crushing food

▶ Flea larvae (young) resemble maggots. Many types of the larvae feed on the dried droppings of adult fleas, and in this way, each generation provides nourishment for the next. After shedding its skin three times, the larva becomes a silk-cocooned pupa, then an adult flea.

Stem of houseplant

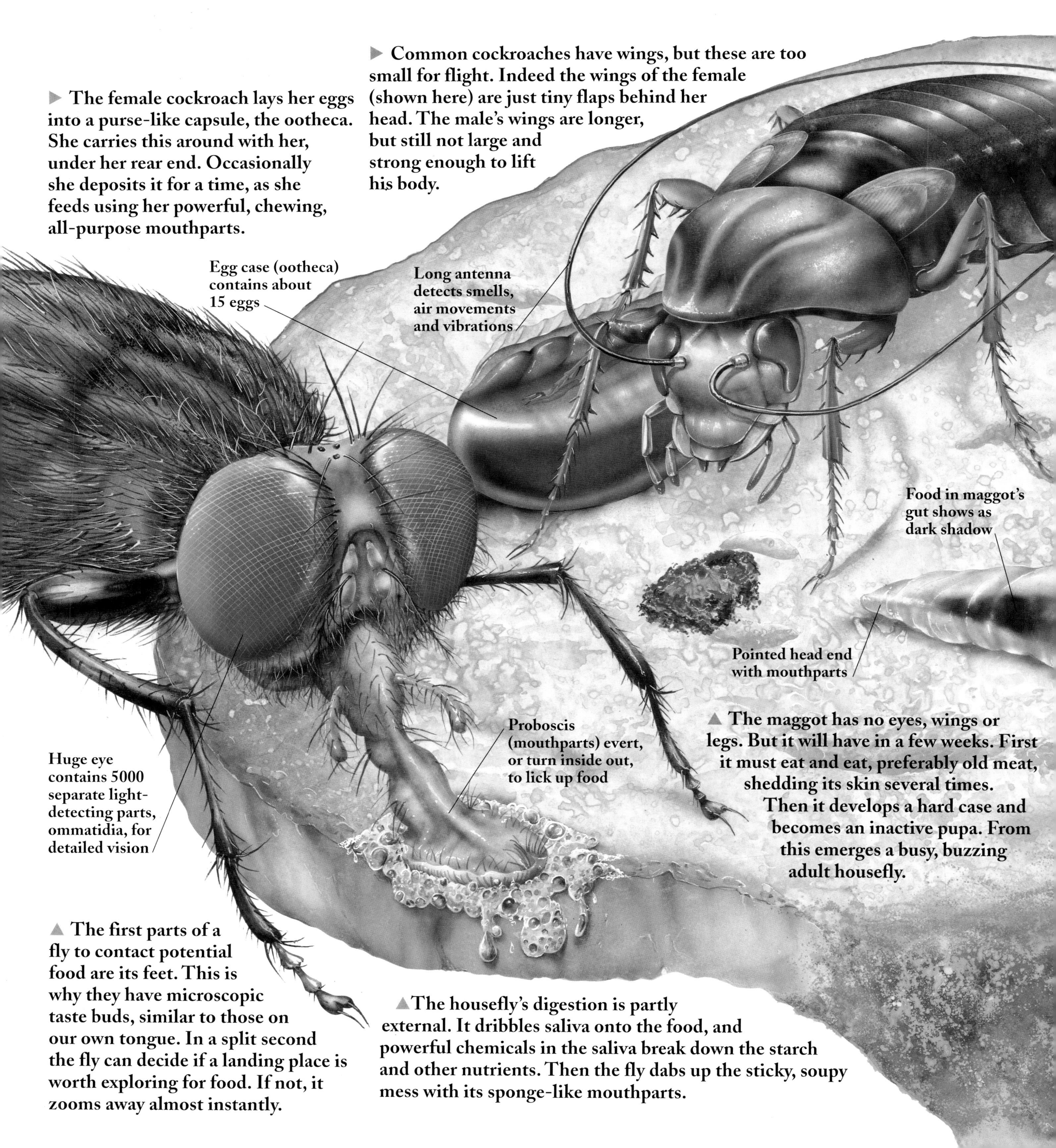
Egg case (ootheca) contains about 15 eggs
Long antenna detects smells, air movements and vibrations
Food in maggot's gut shows as dark shadow
Pointed head end with mouthparts
Proboscis (mouthparts) evert, or turn inside out, to lick up food
Huge eye contains 5000 separate light-detecting parts, ommatidia, for detailed vision

▶ The female cockroach lays her eggs into a purse-like capsule, the ootheca. She carries this around with her, under her rear end. Occasionally she deposits it for a time, as she feeds using her powerful, chewing, all-purpose mouthparts.

▶ Common cockroaches have wings, but these are too small for flight. Indeed the wings of the female (shown here) are just tiny flaps behind her head. The male's wings are longer, but still not large and strong enough to lift his body.

▲ The maggot has no eyes, wings or legs. But it will have in a few weeks. First it must eat and eat, preferably old meat, shedding its skin several times. Then it develops a hard case and becomes an inactive pupa. From this emerges a busy, buzzing adult housefly.

▲ The first parts of a fly to contact potential food are its feet. This is why they have microscopic taste buds, similar to those on our own tongue. In a split second the fly can decide if a landing place is worth exploring for food. If not, it zooms away almost instantly.

▲The housefly's digestion is partly external. It dribbles saliva onto the food, and powerful chemicals in the saliva break down the starch and other nutrients. Then the fly dabs up the sticky, soupy mess with its sponge-like mouthparts.

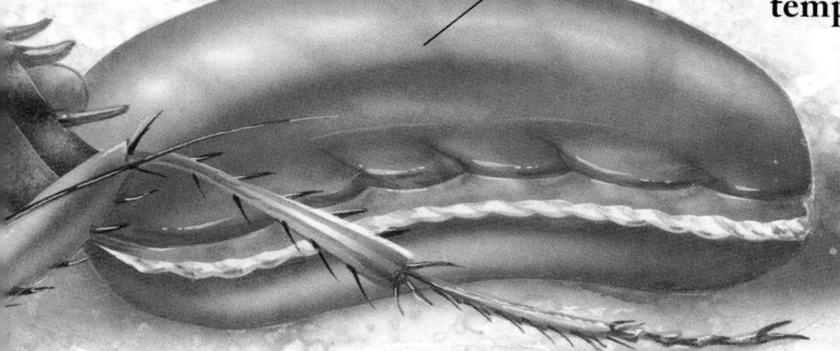

Nymphs (young forms) will hatch from egg case in 2–3 months, depending on temperature

MAGNIFICATION

x25

A slice of old bread feeds hundreds of minibeasts, who do not mind it being stale. The larger diners such as flies and 'roaches moisten the bread with their saliva and digestive juices, to chew off lumps. Then the smaller ones pick away at the damp, softened edges.

Sharing our food

One of the animal world's favourite tastes is starch. It occurs as a natural food store in seeds, intended for the baby plants inside. Too often, however, animals get there first – including us. We grow and harvest seeds, as cereal grains. We mill them to obtain the starch as flour, and bake this into bread. But many other creatures benefit from our efforts, too. They are attracted from far and wide by the smell of bread and the taste of starch. They invade our kitchens, larders and cupboards, to steal our staple foodstuff.

▼ The firebrat, like the silverfish, is a type of wingless insect known as a bristletail or thysanuran. It feeds on starchy food scraps, and shuns light, so is out at night.

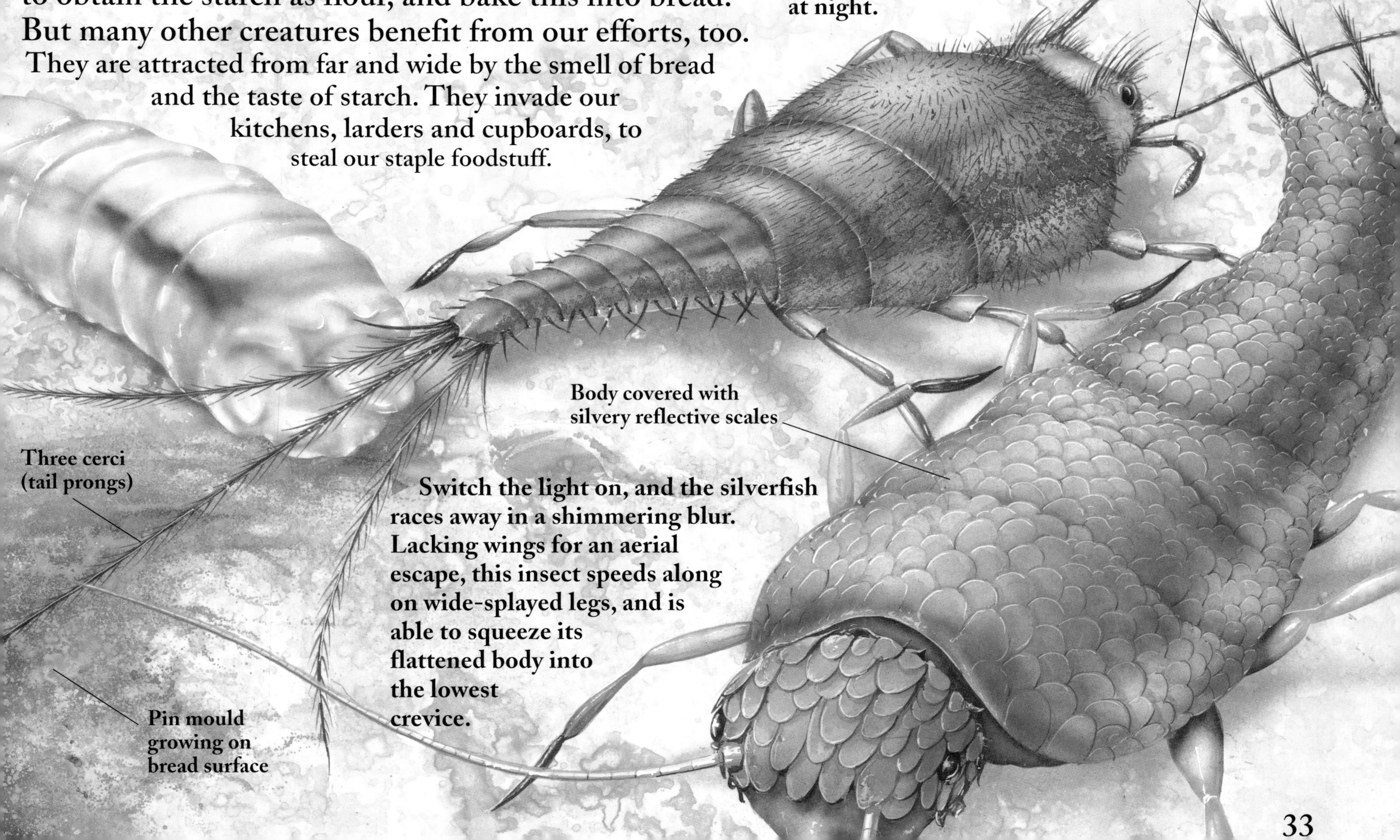

Long and short pairs of antennae

Body covered with silvery reflective scales

Three cerci (tail prongs)

Switch the light on, and the silverfish races away in a shimmering blur. Lacking wings for an aerial escape, this insect speeds along on wide-splayed legs, and is able to squeeze its flattened body into the lowest crevice.

Pin mould growing on bread surface

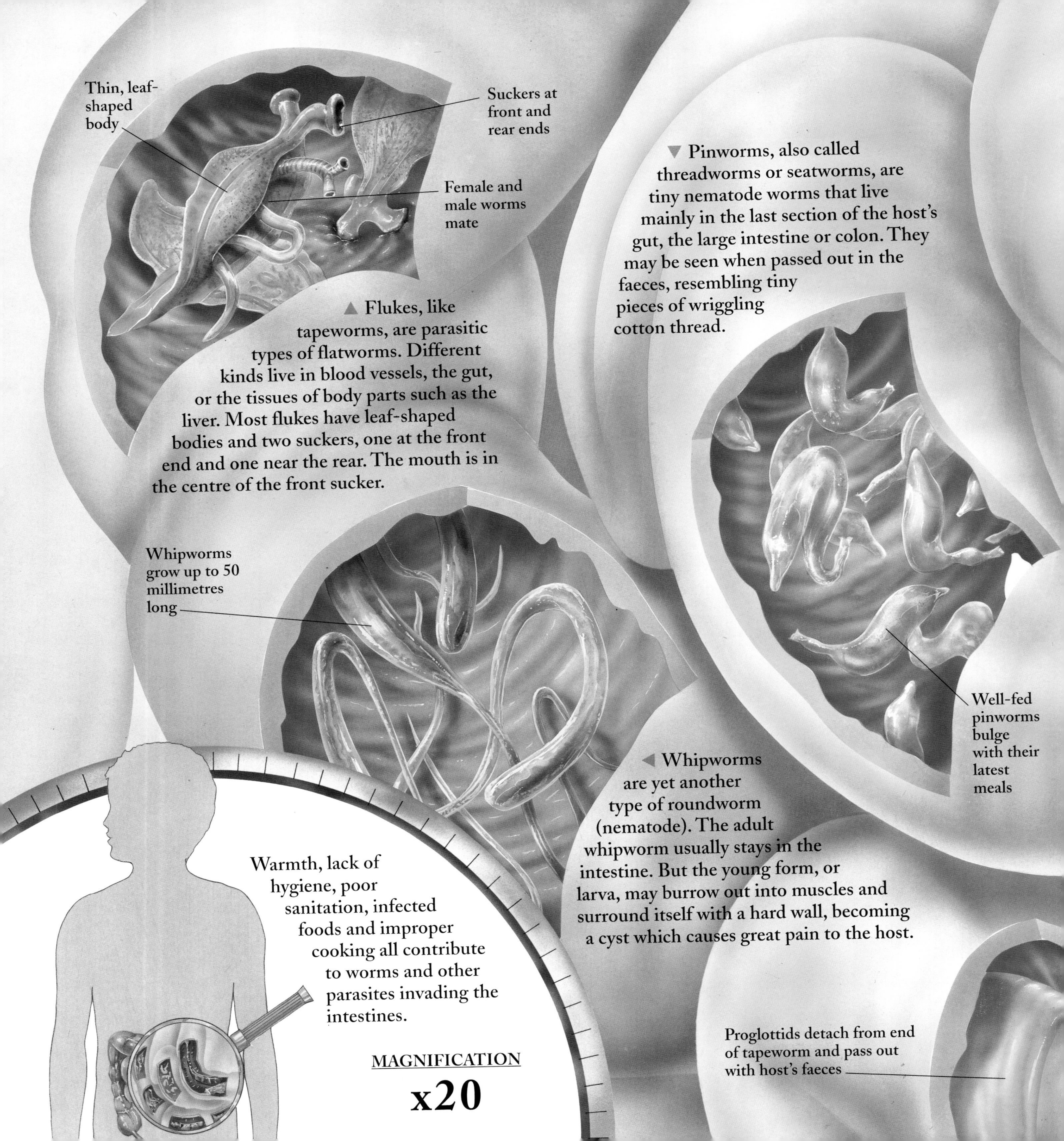
Thin, leaf-shaped body
Suckers at front and rear ends
Female and male worms mate
▲ Flukes, like tapeworms, are parasitic types of flatworms. Different kinds live in blood vessels, the gut, or the tissues of body parts such as the liver. Most flukes have leaf-shaped bodies and two suckers, one at the front end and one near the rear. The mouth is in the centre of the front sucker.
▼ Pinworms, also called threadworms or seatworms, are tiny nematode worms that live mainly in the last section of the host's gut, the large intestine or colon. They may be seen when passed out in the faeces, resembling tiny pieces of wriggling cotton thread.
Whipworms grow up to 50 millimetres long
Well-fed pinworms bulge with their latest meals
◀ Whipworms are yet another type of roundworm (nematode). The adult whipworm usually stays in the intestine. But the young form, or larva, may burrow out into muscles and surround itself with a hard wall, becoming a cyst which causes great pain to the host.
Warmth, lack of hygiene, poor sanitation, infected foods and improper cooking all contribute to worms and other parasites invading the intestines.
Proglottids detach from end of tapeworm and pass out with host's faeces
MAGNIFICATION
x20

A gut-full of pests

Warm, safe from predators, and full of semi-digested food, the gut is a wonderful place to live. Many kinds of true worms, flatworms, nematode worms and other worms thrive here. The tapeworm is not so much a single creature, as a ribbon-shaped colony of individuals – but each one is little more than a few body parts in a bag. Hookworms infest perhaps 1000 million people worldwide, and those in a severely affected person may release 10 million eggs daily. Some of these creatures are remarkably long-lived for their kind. Hookworms can survive for up to 10 years.

Tapeworm head, called the scolex, shown at extra magnification

Hooks

Suckers

Developing proglottids (sections) of tapeworm body

Head of hookworm armed with strong, curving hooks

Body of hookworm has very thin outer layer

Tapeworm is up to ten metres long, folded zig-zag fashion into the six-metre gut

Folded inner surface (mucosa) of gut

Muscular intermediate layers of gut

Tough external coat of gut

Hookworm digests blood, mucus and tissues lining the gut

Maturing proglottids of tapeworm

◀ The hookworm, named after the fierce-looking grab-hooks at its front end, is a type of roundworm or nematode. It may be swallowed by accident as an egg or tiny larva in infected food. Or it may burrow through the host's skin in larval form, be carried by the blood to the lungs, get coughed up the windpipe, and then be swallowed into the gut.

◀ The tapeworm buries its pinhead-sized front end into the lining of the gut, hanging on firmly with hooks and suckers (above). The sections of its long body are not true segments, as in an earthworm. Each is a bag called a proglottid, containing mainly sex organs that produce eggs.

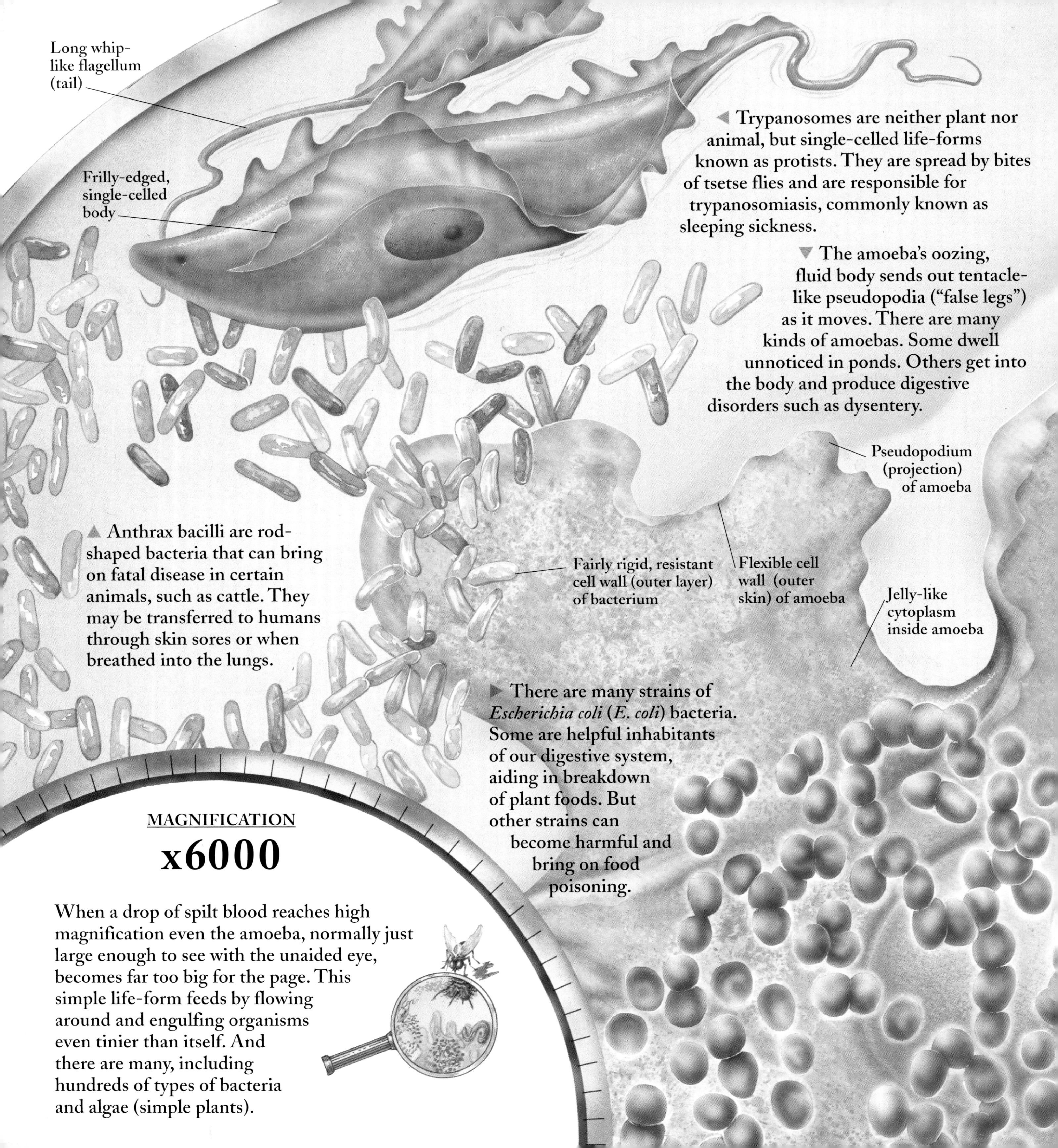

When a drop of spilt blood reaches high magnification even the amoeba, normally just large enough to see with the unaided eye, becomes far too big for the page. This simple life-form feeds by flowing around and engulfing organisms even tinier than itself. And there are many, including hundreds of types of bacteria and algae (simple plants).

Dreaded disease

Blood carries around the body all the ingredients needed for life: oxygen, energy-packed sugars, nutrients, vitamins and minerals. Which is why it's such an inviting liquid for disease-causing microbes and other invaders. They get in through tiny cuts, sores or the thin linings of the nose, throat, lungs or guts. Once within, they thrive and multiply in blood's nutritious warmth – and this is when the troubles may begin.

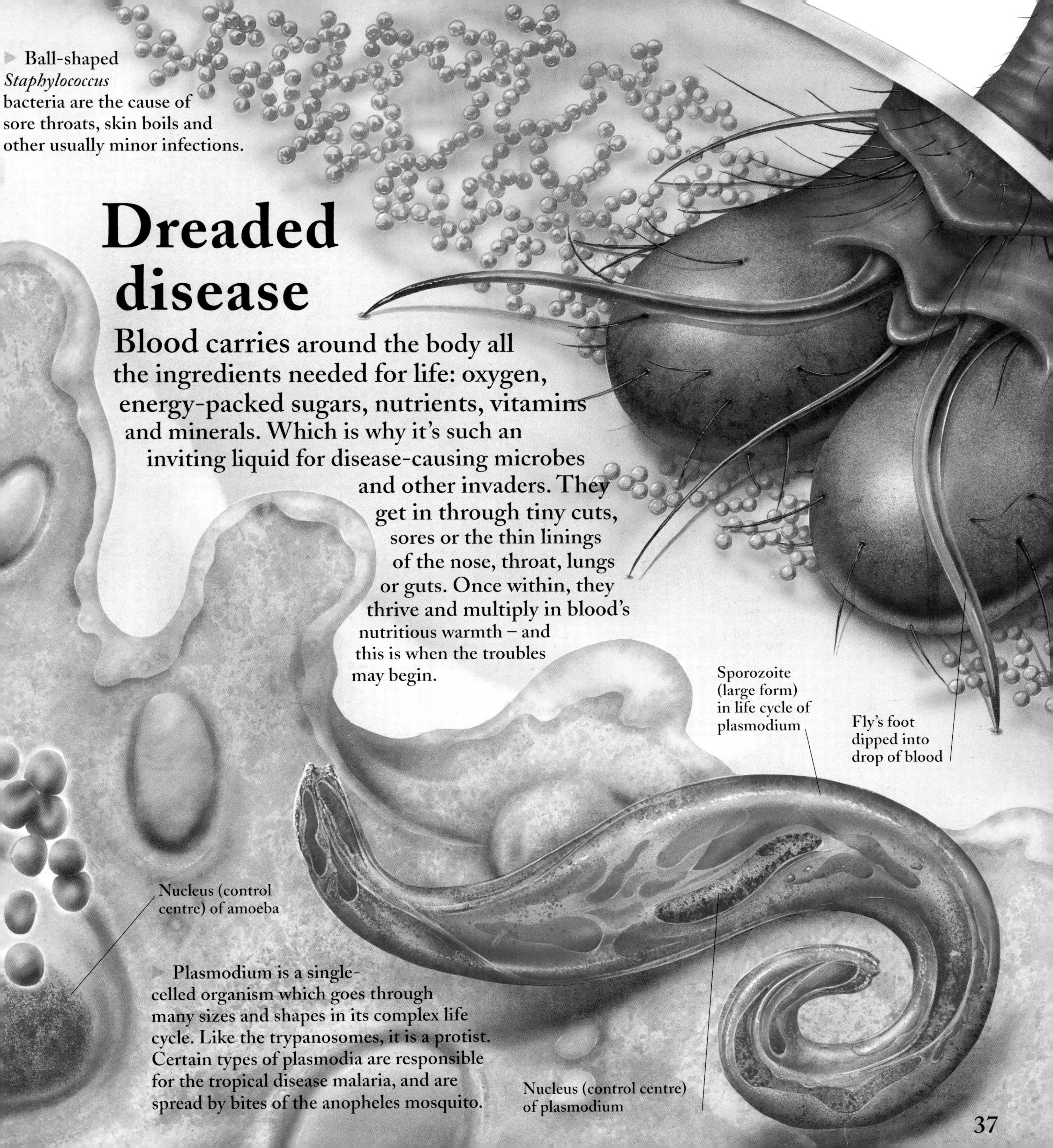

▶ Ball-shaped *Staphylococcus* bacteria are the cause of sore throats, skin boils and other usually minor infections.

▶ Plasmodium is a single-celled organism which goes through many sizes and shapes in its complex life cycle. Like the trypanosomes, it is a protist. Certain types of plasmodia are responsible for the tropical disease malaria, and are spread by bites of the anopheles mosquito.

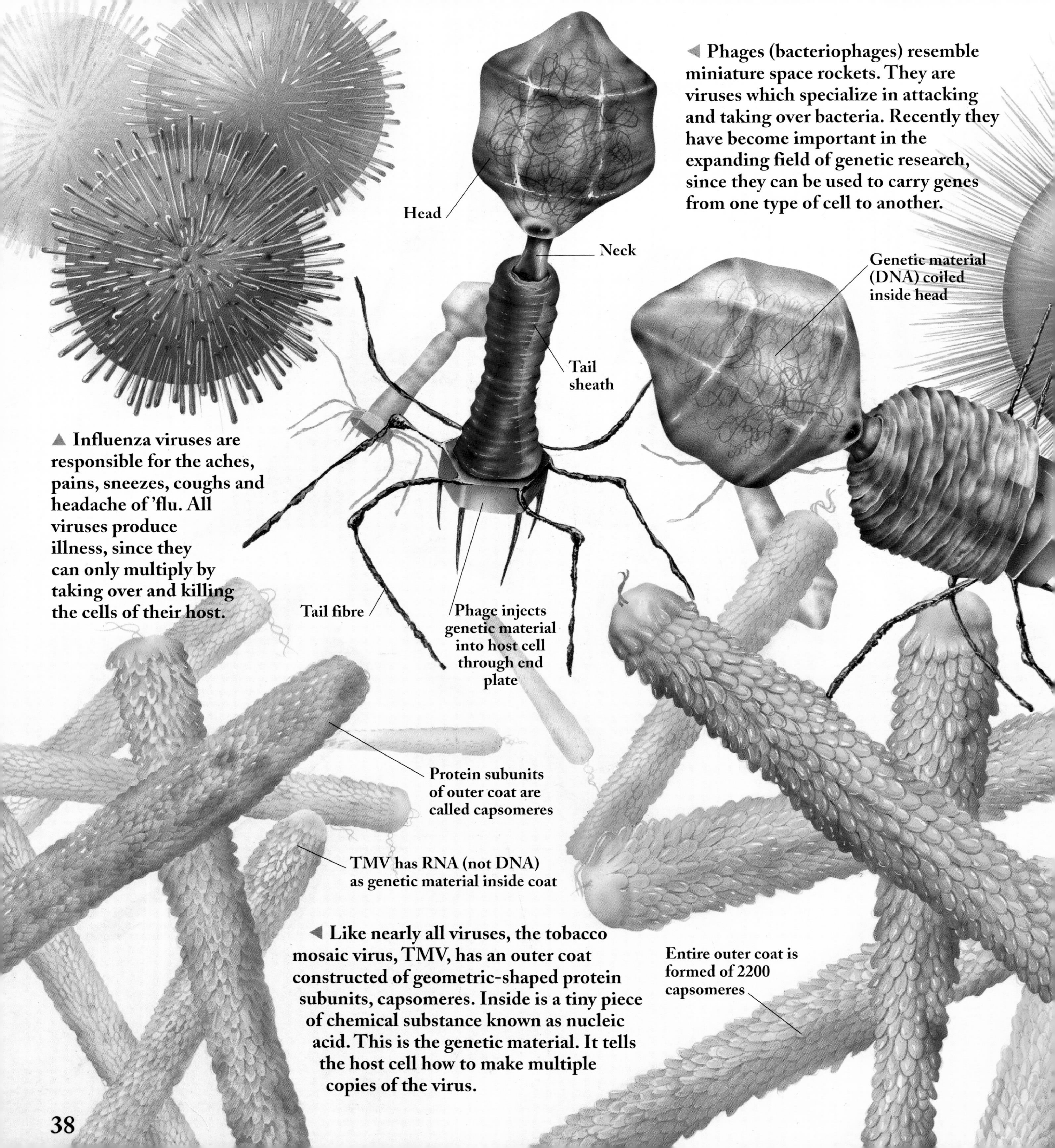

◀ Phages (bacteriophages) resemble miniature space rockets. They are viruses which specialize in attacking and taking over bacteria. Recently they have become important in the expanding field of genetic research, since they can be used to carry genes from one type of cell to another.

▲ Influenza viruses are responsible for the aches, pains, sneezes, coughs and headache of 'flu. All viruses produce illness, since they can only multiply by taking over and killing the cells of their host.

◀ Like nearly all viruses, the tobacco mosaic virus, TMV, has an outer coat constructed of geometric-shaped protein subunits, capsomeres. Inside is a tiny piece of chemical substance known as nucleic acid. This is the genetic material. It tells the host cell how to make multiple copies of the virus.

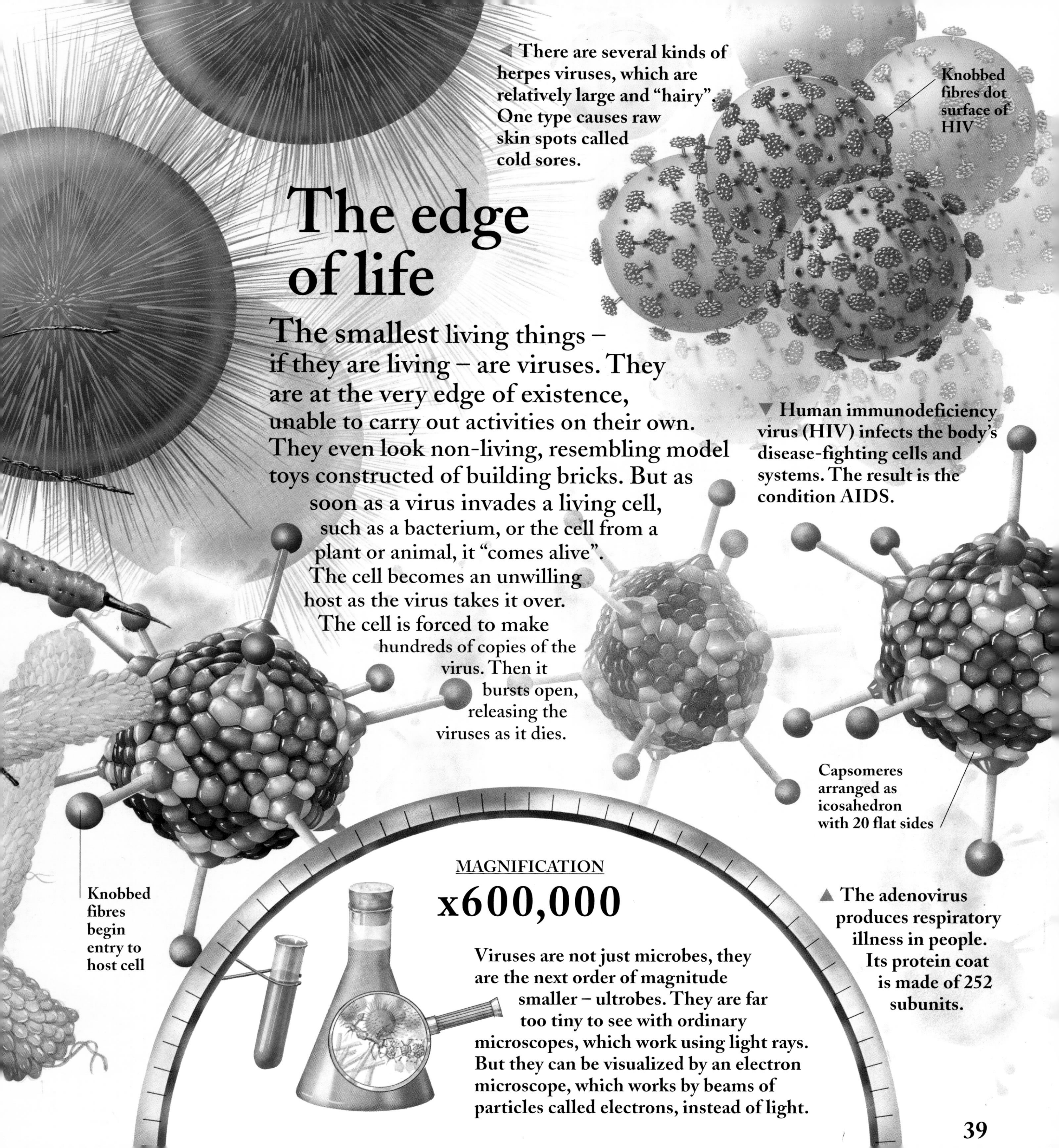

◀ There are several kinds of herpes viruses, which are relatively large and "hairy". One type causes raw skin spots called cold sores.

Knobbed fibres dot surface of HIV

The edge of life

The smallest living things – if they are living – are viruses. They are at the very edge of existence, unable to carry out activities on their own. They even look non-living, resembling model toys constructed of building bricks. But as soon as a virus invades a living cell, such as a bacterium, or the cell from a plant or animal, it "comes alive". The cell becomes an unwilling host as the virus takes it over. The cell is forced to make hundreds of copies of the virus. Then it bursts open, releasing the viruses as it dies.

▼ Human immunodeficiency virus (HIV) infects the body's disease-fighting cells and systems. The result is the condition AIDS.

Capsomeres arranged as icosahedron with 20 flat sides

Knobbed fibres begin entry to host cell

MAGNIFICATION

x600,000

Viruses are not just microbes, they are the next order of magnitude smaller – ultrobes. They are far too tiny to see with ordinary microscopes, which work using light rays. But they can be visualized by an electron microscope, which works by beams of particles called electrons, instead of light.

▲ The adenovirus produces respiratory illness in people. Its protein coat is made of 252 subunits.

Micro-files

This section provides descriptions and details of the living things shown on previous pages. They are listed not by size, but in their natural groupings. Some of these are familiar, such as Plants, Insects and Fish. Others are less so. The Protists include a huge variety of life-forms with the common feature that each is a single microscopic cell. And there are at least three major groups of worms!

VIRUSES

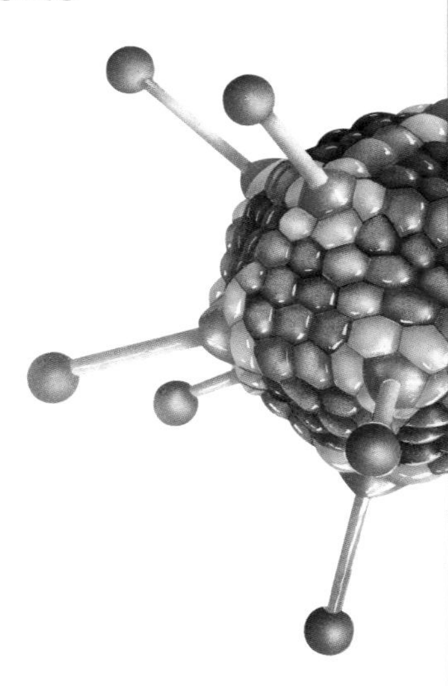

Adenovirus (p 39)
Adenoviridae
• SIZE RANGE 0.00008 mm
Found in the respiratory systems of animals and people, these viruses can cause various infections such as colds and sore throats, with coughing, sneezing and similar symptoms.

Herpes virus (p 38–39)
Herpesviridae
• SIZE RANGE 0.00015 mm
Various kinds of herpes viruses infect animals and people, but may not cause obvious symptoms until they are activated by some event, such as another illness. They are involved in cold sores, gum disease, chickenpox and certain tumours.

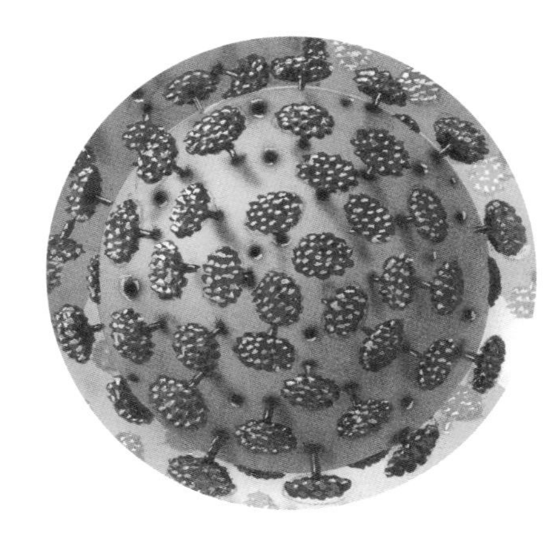

HIV (p 39)
(Human immunodeficiency virus)
• SIZE RANGE 0.000075 mm
HIV is the cause of the condition AIDS. It attacks certain body cells that are involved in the workings of the immune system, which protects the body from disease. This means that the immunity of the body is lowered and the resistance to diseases is weakened.

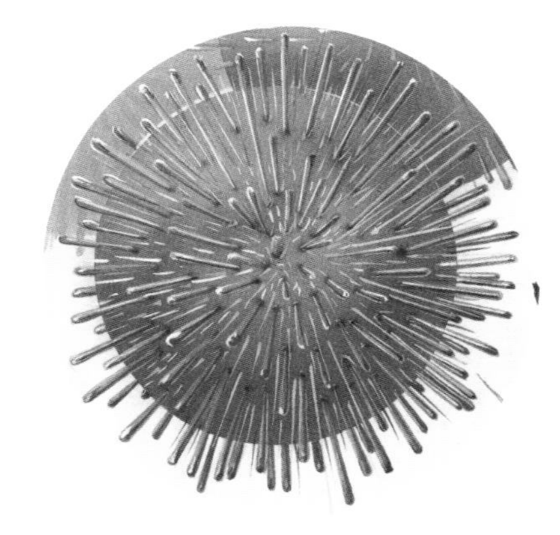

Influenza virus (p 38)
Orthomyxovirus (various types)
• SIZE RANGE 0.0001 mm
A liking for the slimy mucus which lines the passages of the nose, throat, windpipe and lungs means that the 'flu virus causes infection of the respiratory system, with sneezes, coughs, also fever and headache.

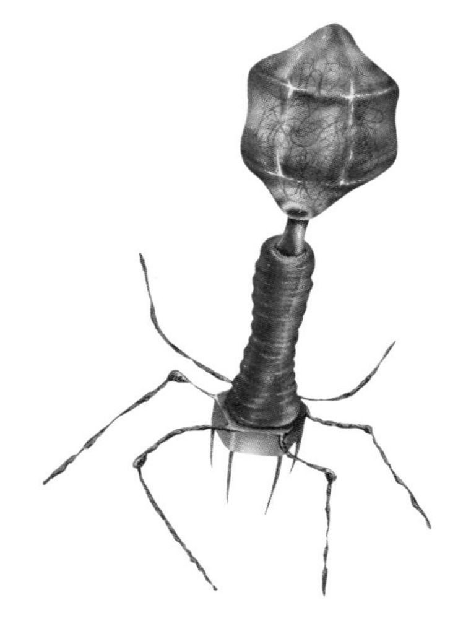

Phage (p 38–39)
Bacteriophage T4 virus and others
• SIZE RANGE 0.00025 mm
Phages or bacteriophages are viruses that attack and invade bacteria. There are types of phage for each kind or strain of bacteria. Phages are also used in genetics to carry genes from one cell to another.

Tobacco mosaic virus (p 38–39)
• SIZE RANGE Length 0.0003 mm
This virus produces disease in plants such as tobacco, tomato and potato. The leaves become misshapen and develop a patchwork or mosaic of light green and dark green areas.

BACTERIA

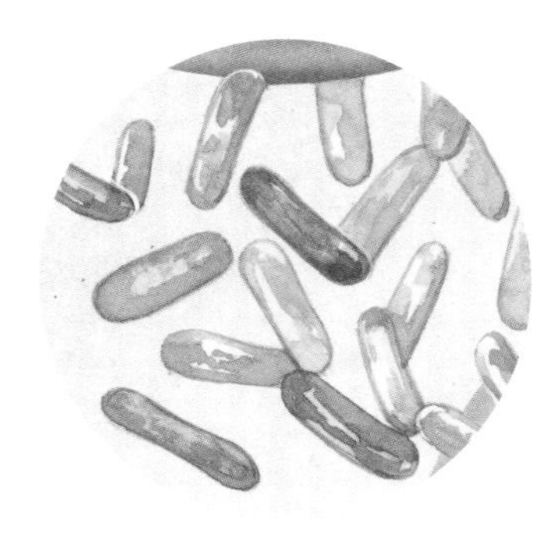

Anthrax (p 36)
Bacillus anthracis
• SIZE RANGE 0.003 mm
• HOW IT FEEDS **Absorbs nutrients from the fluids around it.**
Anthrax bacteria cause the disease of the same name in cattle, sheep and other farm animals. This can also affect people, causing severe sores or pustules, and also a type of pneumonia.

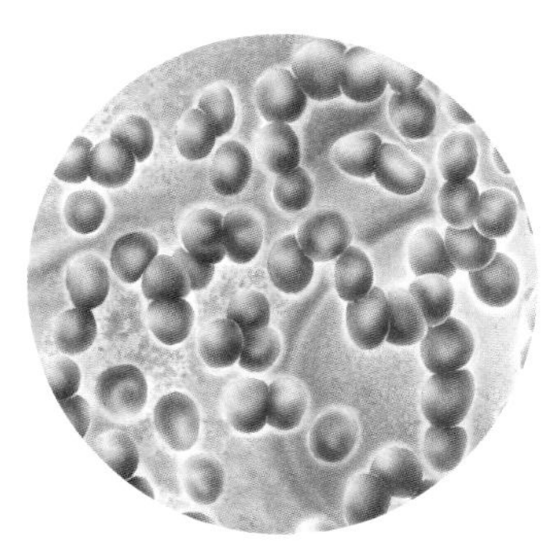

E. coli (p 36–37)
Escherichia coli
• SIZE RANGE About 0.002 mm
• HOW IT FEEDS **Absorbs nutrients from the fluids around it.**
E. coli are very common bacteria, found in a variety of places. Some types occur in the human intestine, where they are helpful rather than harmful, assisting digestion. Other strains cause illness.

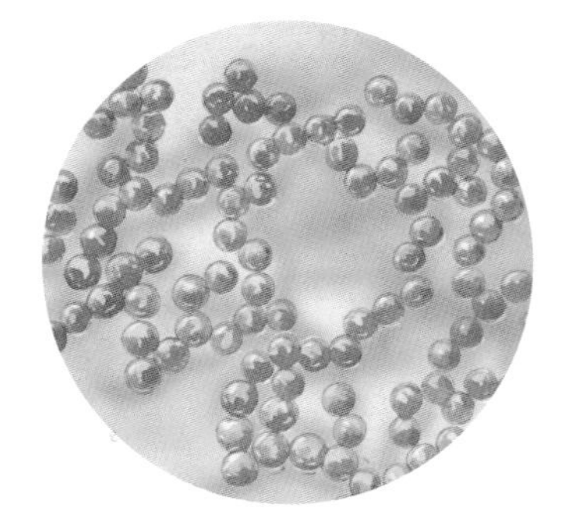

Staphylococcus (p 37)
S. albus, aureus (pyogenes) and others
• SIZE RANGE About 0.001 mm
• HOW IT FEEDS **Absorbs nutrients from the fluids around it.**
Some types of these bacteria live on skin, where they cause few problems, perhaps small spots and pimples. More harmful types cause boils, sores, abscesses and infections of the nose, throat and urinary system.

PROTISTS

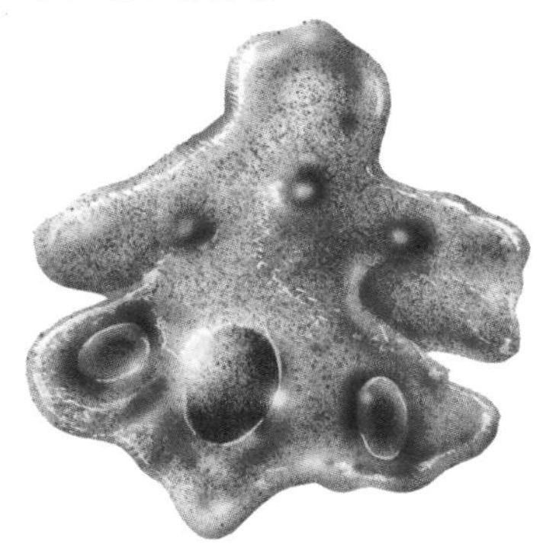

Amoeba (p 36–37)
Amoeba (various kinds), *Rhizopoda*
• SIZE RANGE **0.01–0.1 mm**
• HOW IT FEEDS **Flows around and engulfs tiny food particles.**
There are many kinds of amoeba. Some are free-living in ponds, streams, soil and salty water. Others are parasites inside animals and plants. *Entamoeba histolytica* causes amoebic dysentery in people.

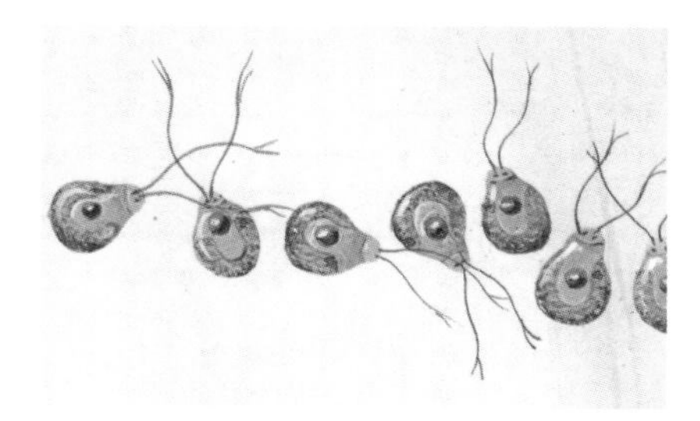

Chlamydomonas (p 12)
Phytoflagellate
• SIZE RANGE **Less than 1 mm**
• HOW IT FEEDS **Absorbs light energy from the sun by the process of photosynthesis.**
Chlamydomonas is one of thousands of kinds of single-celled organism, which resemble plants in their ability to trap light energy using green pigments. They swim by waving long filaments called flagella.

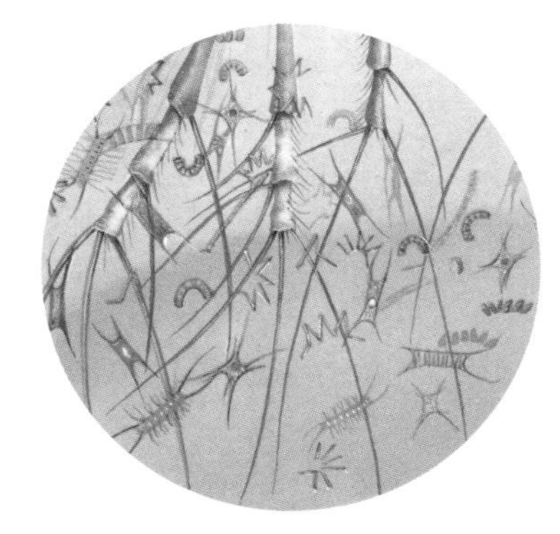

Diatom (p 11, 13)
Bacillariophyceae (*Algae*)
• SIZE RANGE **Mostly 1 mm or less**
• HOW IT FEEDS **Traps light energy from the sun by photosynthesis.**
Diatoms are variously regarded as very simple plants, each of just a single cell, or types of protists. They are a major part of the phytoplankton – the tiny plants which thrive almost everywhere in seawater.

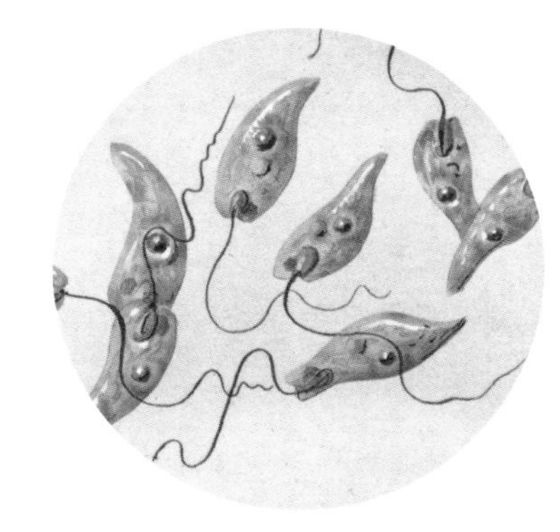

Euglena (p 12)
Phytoflagellate
• SIZE RANGE **Less than 1 mm**
• HOW IT FEEDS **Absorbs light energy from the sun, and engulfs food particles.**
Euglena is an example of one of the thousands of kinds of protist, which can gain energy either from sunlight, like a plant, or "eat" food like an animal. To swim, it waves its long filament, or flagellum.

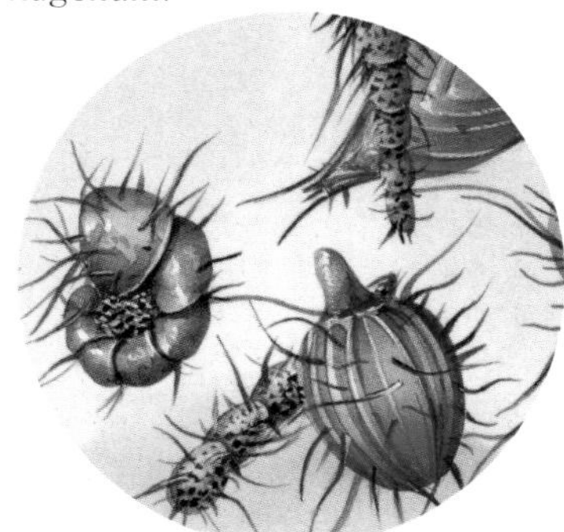

Foraminiferan (p 10)
Foraminiferida (*Rhizopoda/Sarcodina*)
• SIZE RANGE **Diameter up to 1 mm**
• HOW IT FEEDS **Consumes or engulfs microscopic particles such as bacteria.**
Called "forams" for short, these protists are related to the amoeba. Each has a tough case, or test, made of hard minerals. Test shapes vary enormously. Some are multi-chambered and beautifully patterned.

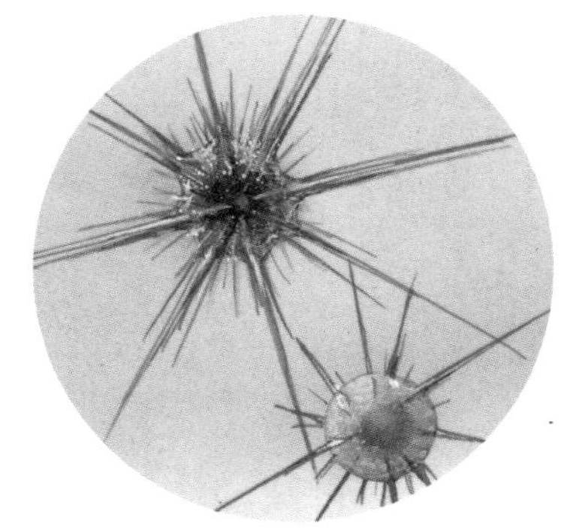

Heliozoan (p 10)
Heliozoia (*Sarcodina*)
• SIZE RANGE **Diameter up to 1 mm**
• HOW IT FEEDS **Consumes or engulfs microscopic particles such as bacteria.**
"Heliozoans" means "sun-animals", from their resemblance to a shining sun. Each of these protists has a central cell mass surrounded by long, slim, tapering projections called axopodia.

Plasmodium (p 37)
P. falciparum and others
• SIZE RANGE **Up to 0.05 mm**
• HOW IT FEEDS **Absorbs nutrients from the fluids around it.**
Spread by mosquitoes, these single-celled parasites cause malaria. They go through many stages in their life cycle, producing the main symptoms of fever, headache and shivering when they attack red cells in the host's blood.

Trypanosome (p 36)
Trypanozoon / Trypanosoma
• SIZE RANGE **About 0.03 mm**
• HOW IT FEEDS **Absorbs nutrients from the fluids around it.**
These single-celled parasites cause various diseases, including two kinds of sleeping sickness mainly in Africa (spread by biting flies), and Chagas' disease which occurs chiefly in South America.

FUNGI

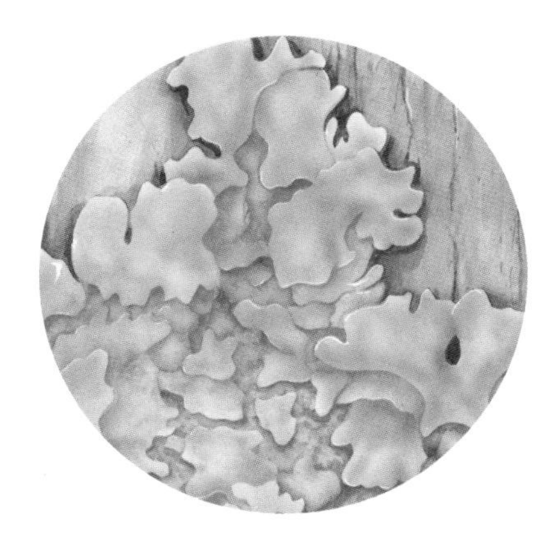

Lichen (p 18)
Cladocera and many others
• SIZE RANGE **Grow as crusts or moss-like patches**
• HOW IT FEEDS **Absorbs light energy, takes in simple nutrients.**
Lichens are mutually helpful partnerships of very small, simple, green plants called algae, and various kinds of moulds (fungi). They grow almost everywhere, including bark, stones and rocks, but dislike pollution.

Mushroom (p 20–21)
Agaricus and many others
• SIZE RANGE **Diameter 1–10 cm**
• HOW IT FEEDS **Digests and absorbs rotting matter, sap and juices.**
The umbrella-shaped mushroom cap, on its stalk, is only the "fruiting body" (reproductive part) of the whole fungal organism. It releases the spores. The rest of the organism is a maze of underground threads.

Penicillium mould (p 23)
Penicillium
• SIZE RANGE **Variable patches**
• HOW IT FEEDS **Digests and absorbs rotting, decaying matter.**
Many kinds of *Penicillium* and similar moulds grow as fluffy or hairy patches on old fruit, in soil, and wherever there is organic matter. The original antibiotic drug, penicillin, was obtained from a strain of this mould.

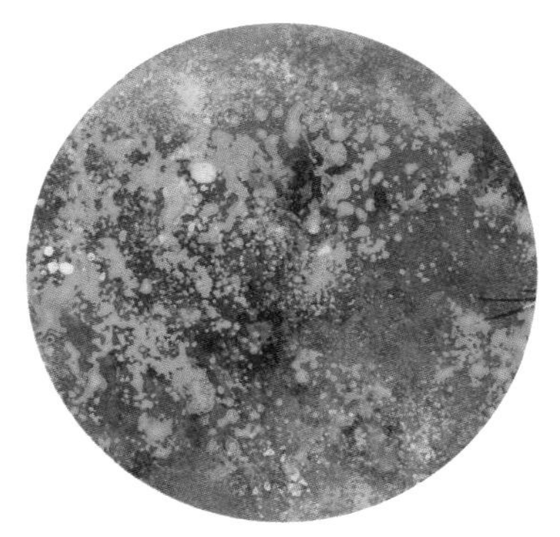

Pin mould (p 32–33)
Mucor
• SIZE RANGE **Forms variable patches**
• HOW IT FEEDS **Produces digestive juices and absorbs a kind of "soup".**
Pin mould is one of many kinds of whitish mould on old bread, decaying fruit, stale vegetables and similar items. It produces tiny, rounded, spore-releasing fruiting bodies on stalks, that look like miniature pins.

Slime mould (p 20)
Myxomycetes
• SIZE RANGE **Mostly 1-5 cm**
• HOW IT FEEDS **Digests and absorbs rotting matter, sap and juices.**
Slime moulds are jelly-like masses with no fixed shape, that creep about on rotting wood. Gradually the mass stops growing and moving to produce stalks which release spores.

PLANTS

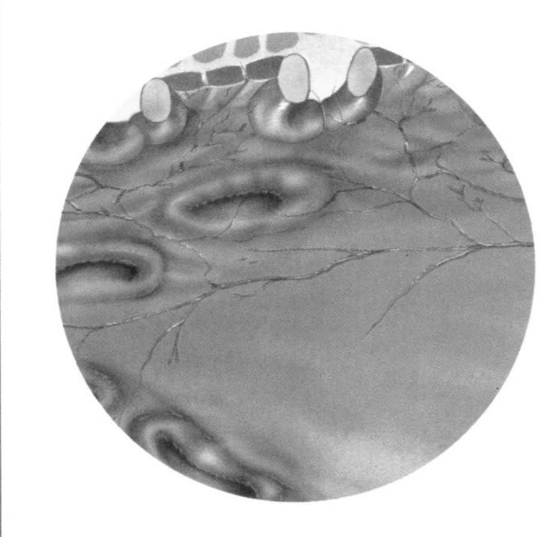

Leaf blade (p 14–15)
Lamina
• SIZE RANGE **Depends on plant**
• WHAT THEY ARE **The usually flat, green parts of a plant, specialized to absorb light energy.**
Leaves are nature's solar-powered factories. They absorb light energy by photosynthesis and use it to join together simple nutrients, to make high-energy sugars and starches that power the plant's life processes.

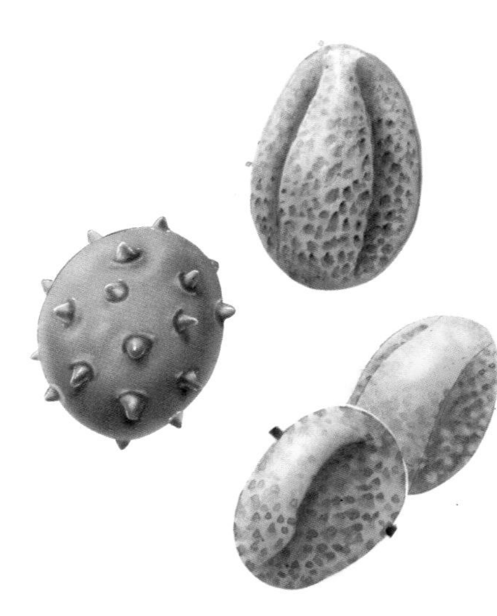

Orchid seeds, Poppy seed (pages 9, 23)
Orchidaceae, Papaveraceae
• SIZE RANGE **Diameter up to 1 mm**
• WHAT THEY ARE **Tiny seeds, among the smallest of any plant seeds.**
These seeds are extremely small and light, designed for dispersal away from the parent plant by the wind. There are more than 17,500 natural species of orchids, and even more cultivated varieties.

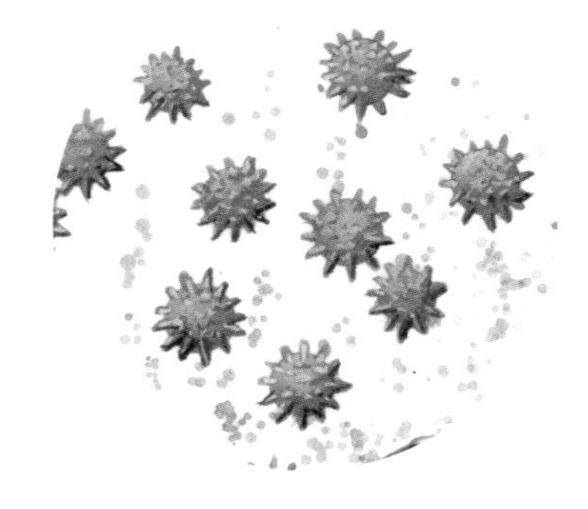

Pollen (p 8, 16)
Produced by most flowers
• SIZE RANGE **Up to 1 mm across**
• WHAT IT IS **Male reproductive cells of plants encased in tough skins.**
Pollen grains are mostly tiny, pale objects that resemble yellow dust. They are carried from the anthers (male parts) of a flower to the carpels (female parts) of another flower of the same kind, so that seeds can develop.

Stinging nettle (p 14)
Urtica
• SIZE RANGE **Height 1 m or more**
• HOW IT FEEDS **Absorbs light energy from the sun, and minerals and nutrients through its roots.**
There are about 30 kinds of stinging nettle. Most have small greenish flowers and are covered with small hairs that snap to release irritant, stinging chemicals. This is a defence against herbivores.

TARDIGRADA

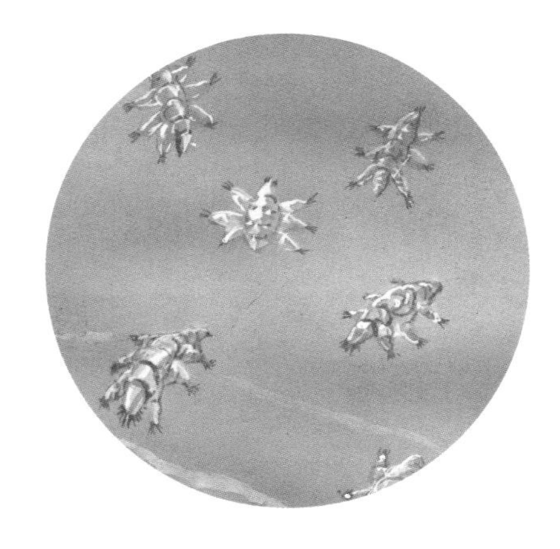

Tardigrade (p 26)
Tardigrada
• SIZE RANGE **Length less than 1 mm**
• HOW IT FEEDS **Most suck up plant juices, some are predators.**
"Tardigrade" means "slow stepper" and this is how the animal moves, walking slowly and deliberately like a tiny caterpillar on its four pairs of stumpy legs. These creatures are also called water-bears or moss-bears.

ROTIFERA

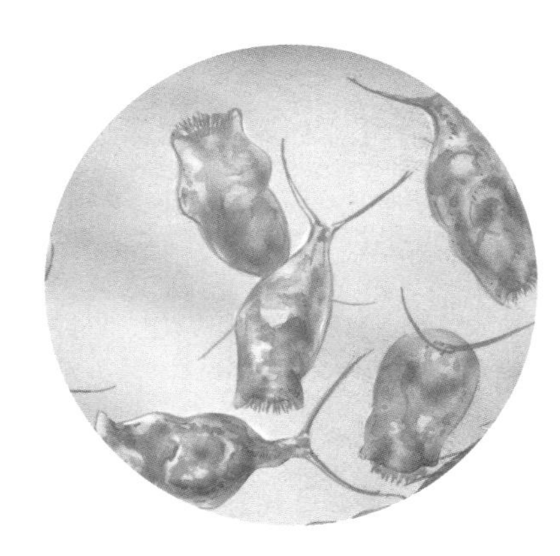

Rotifer (p 27)
Rotifera
• SIZE RANGE **Less than 1 mm across**
• HOW IT FEEDS **Filters microscopic edible particles from water.**
There are more than 2000 different kinds of these "wheel-animalcules" but most are smaller than pinheads. They are cup-shaped with a row of hairs around the rim (the head end) and slim tail stalks.

CNIDARIA

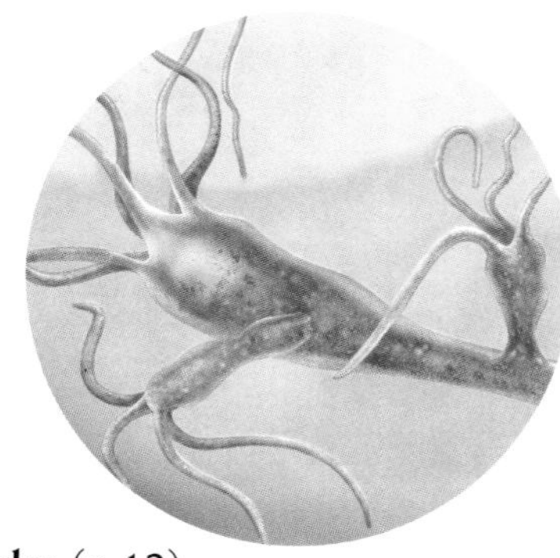

Hydra (p 12)
Hydroida
• SIZE RANGE **Length up to 10 cm**
• HOW IT FEEDS **Grasps tiny food with its flexible, stinging tentacles.**
Most hydras in ponds and streams grow to just a few millimetres long, but occasional giants occur. They resemble tall, thin, tree-shaped versions of their close cousins, sea anemones, and follow a similar lifestyle.

WORMS

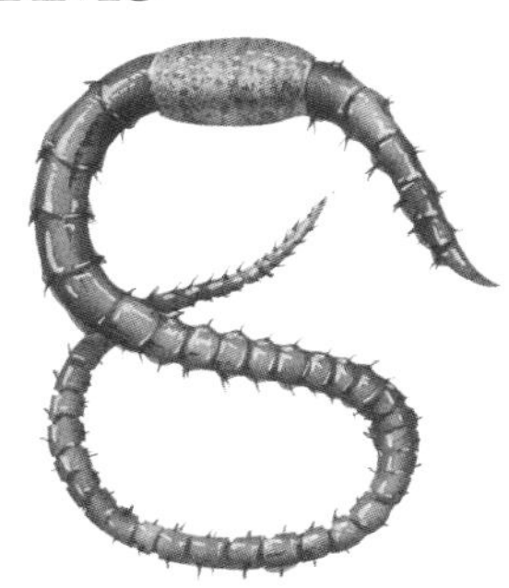

Bloodworm (p 26)
Tubifex and others
• SIZE RANGE **Length up to 5 cm**
• HOW IT FEEDS **Eats mud and digests any edible particles.**
Also called sludgeworms, these bright red worms are common in still and stagnant water. They live part-buried in burrows in the mud and form writhing, undulating, reddish masses on the bottom.

Earthworm (p 22)
Lumbricus and many others
• SIZE RANGE **Mostly less than 30 cm**
• HOW IT FEEDS **Eats and digests any edible particles in soil.**
Earthworms wriggle and munch through soil, helping to recycle organic material back into minerals and nutrients which plants can take up for growth. Their activities aerate and irrigate the soil and mix its layers.

FLATWORMS

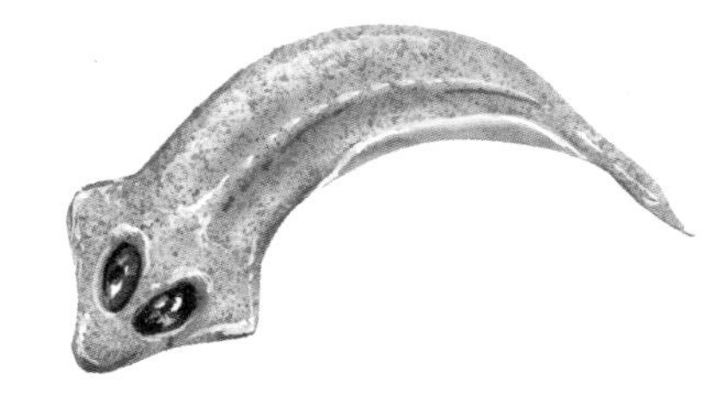

Flatworm (free-living) (p 13, 26–27)
Planaria
• SIZE RANGE **Mostly less than 10 mm long**
• HOW IT FEEDS **Eats tiny prey with the mouth on the underside.**
Free-living (that is, not parasitic) flatworms thrive in the sea, freshwater and damp places. They usually move by gliding along like slugs, but some can undulate their leaf-like bodies and swim. Most are predators.

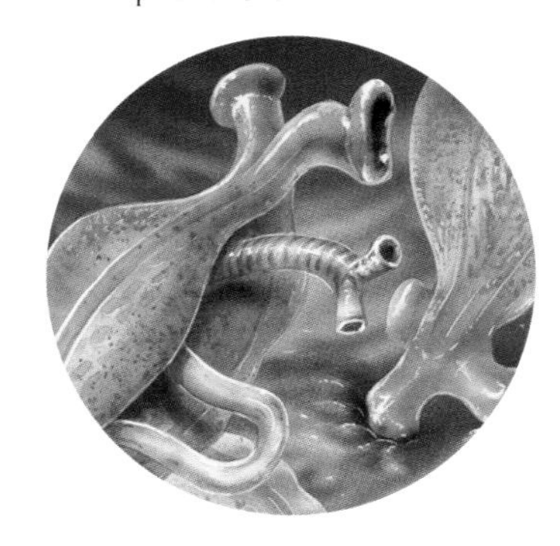

Fluke (p 34)
Schistosoma, Fasciola and many others (*Trematoda*)
• SIZE RANGE **Up to 50 mm**
• HOW IT FEEDS **Sucks blood and absorbs nutrients through its skin.**
Flukes are leaf-shaped, parasitic types of flatworm. There are many kinds, each with a specific host – or often, more than one, since many flukes infest different animals at different stages of their lifecycle.

Tapeworm (p 34–35)
Taenia and others (Cestoda)
• SIZE RANGE **Length up to 10 m**
• HOW IT FEEDS **Absorbs the semi-digested nutrients all around it.**
The tapeworm is not strictly a single creature, but a colony of individuals, each resembling one section or segment. Like most other worm parasites, they can be killed by modern medical drugs.

ROUNDWORMS

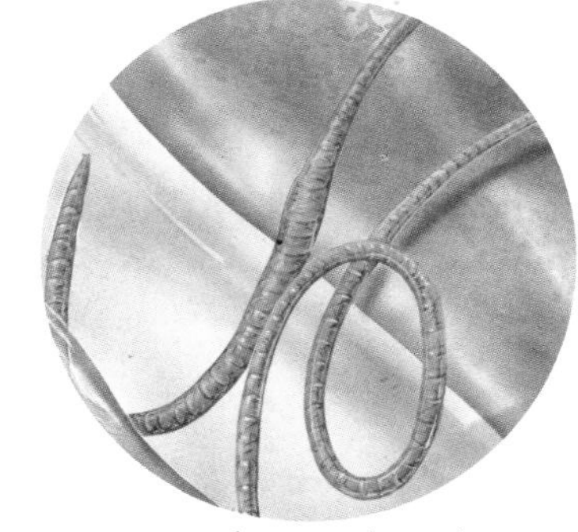

Aquatic roundworm (p 13)
Trilobus, Turbatrix and others
• SIZE RANGE **Length less than 5 mm**
• HOW IT FEEDS **Some eat decaying vegetation, others are predators.**
Various roundworms flourish in all kinds of fresh water, from puddles to fast-flowing streams, and especially stagnant ponds and ditches. Some are commonly known as eel-worms, others as vinegar-worms.

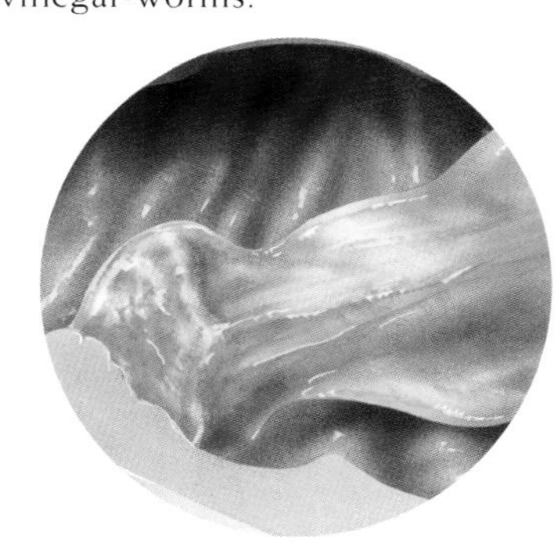

Hookworm (p 35)
Ankylostoma, Necator and others
• SIZE RANGE **Length up to 10 mm**
• HOW IT FEEDS **Eats and absorbs the semi-digested nutrients around it.**
Like many parasitic worms, the hookworm has a complicated lifecycle. Its eggs or larvae are found in soil contaminated with faeces. The adults cause usually slight but persistent blood loss and ill-health.

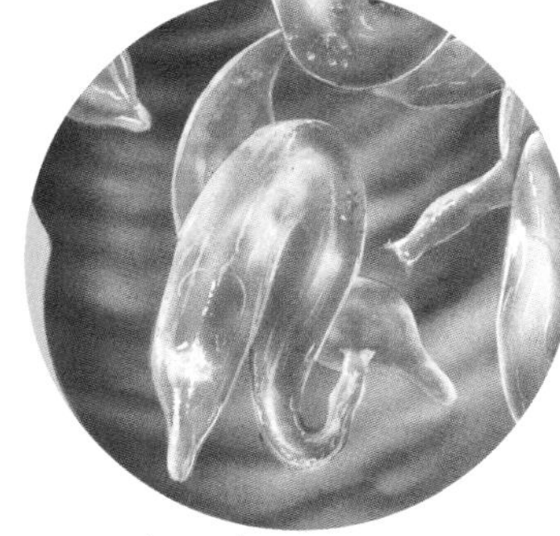

Pinworm (p 34)
Enterobius
• SIZE RANGE **Length up to 10 mm**
• HOW IT FEEDS **Eats contents in, and lining of, the large bowel.**
Pinworms are a gut infection sometimes found in children. They are not usually a major health hazard unless their numbers are very high in one host. But they can cause irritation and itching around the anus.

Soil nematode (p 23, 25)
Trilobidae, Plectidae
• SIZE RANGE **Less than 5 mm long**
• HOW IT FEEDS **Some eat decaying humus and vegetation, others are predators.**
Some eat decaying humus and vegetation, others are predators. Most soil nematodes live in the thin film of water that coats the mineral particles of even the driest-looking earth. As with most other soil dwellers, about nine-tenths of their numbers occur within the uppermost 15 cm.

Whipworm (p 34)
Trichuris
• SIZE RANGE **Length up to 50 mm**
• HOW IT FEEDS **Eats and absorbs the contents of the gut all around it.**
Whipworms are found mainly in warmer areas of the world and infest the intestines of people and other mammals. They cause vague ill-health and loss of blood, but a tangle of numerous worms may block the gut.

ECHINODERMS

Starfish (p 11)
Asteroidea
• SIZE RANGE **Diameter up to 1 m**
• HOW IT FEEDS **Most starfish are predators, digesting and absorbing the flesh of their prey.**
Starfish pass through many strange-shaped larval stages during their early growth. However, all have the radial or circular body design of the adults, often with five spoke-like arms or multiples of five arms.

INSECTS

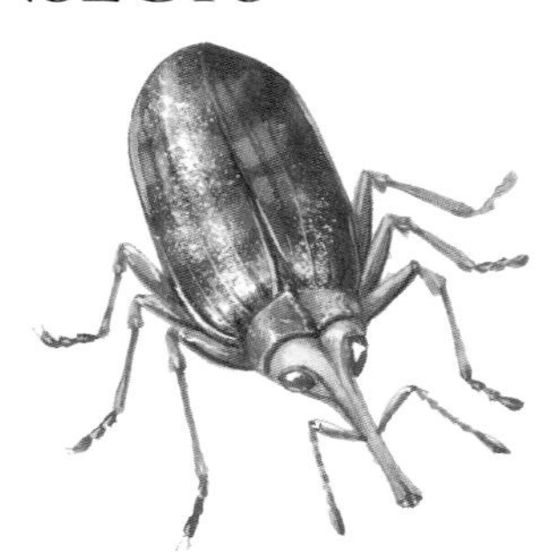

Bark weevil (p 19)
Circulionidae (Coleoptera)
• SIZE RANGE **Length up to 4 mm**
• HOW IT FEEDS **Chews plant matter including leaves, bark, sap wood.**
Weevils are long-snouted cousins of beetles, with hard, domed coverings to their wings. There are many tens of thousands of different kinds, making up one of the largest families in the animal world.

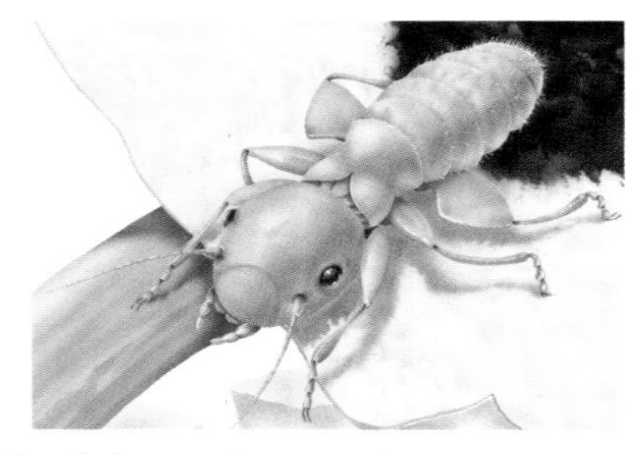

Book louse (page 31)
Psocoptera
• SIZE RANGE **Mostly less than 5 mm**
• HOW IT FEEDS **Bites, scrapes and chews almost any edible fragments and bits.**
Psocopteran lice, unlike "sucking lice", are non-parasitic. They are squat, round-headed, rapid runners and consume all kinds of plant and animal matter, decaying debris, lichens and moulds.

Carpet beetle (p 30)
Dermestidae (Coleoptera)
• SIZE RANGE **Length 1–4 cm**
• HOW IT FEEDS **Larvae eat bits of skin, fur, hair, wool and similar matter.**
The hairy larvae of the carpet or skin beetles move in a jerky fashion. They do most damage to natural fibres, and also attack stored animal products, from woollen rugs to museum collections.

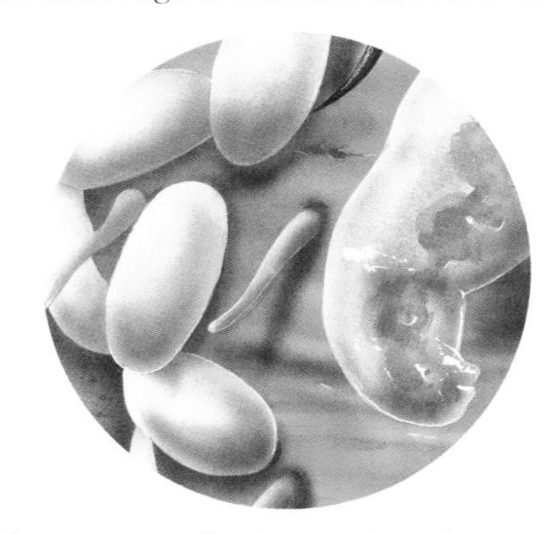

Carrot root-fly (eggs) (p 25)
Psila (Diptera)
• SIZE RANGE **1–2 mm**
• HOW IT FEEDS **Eats the fleshy roots of carrots, also parsnips, parsley and other plants.**
These maggots can become crop pests as they chomp away at juicy roots, causing stunted, misshapen carrots. They change into pupae in the soil, and the adult carrot flies which emerge resemble gnats or midges.

Cockroach (page 32)
Blatta and others *(Blattodea)*
• SIZE RANGE **Length up to 25 mm**
• HOW IT FEEDS **Chews most substances, including our own food scraps.**
The common cockroach cannot fly, although it can run fast and hide in narrow crevices. Several other types of cockroach also infest larders, kitchens, food stores and other buildings, and some of these can fly.

Damselfly (nymph) (p 27)
Zygoptera (Odonata)
• SIZE RANGE **Wingspan 5–10 cm**
• HOW IT FEEDS **Both young and adults are predators of smaller creatures.**
Smaller and more delicate than their close cousins, dragonflies, damselflies are similarly fierce hunters. The nymphs (larval forms) dwell underwater, perhaps for several years.

Firebrat (p 33)
Thermobia (Thysanura)
• SIZE RANGE **Length up to 20 mm**
• HOW IT FEEDS **Munches scraps of spilt and leftover food.**
Firebrats, which are types of three-pronged bristletails, need warm conditions to survive, as in often-used kitchens. They emerge at night to run over floors and worktops, looking for tiny scraps of food.

Flea (larva) (p 31)
Pulex and others *(Siphonaptera)*
• SIZE RANGE **Mostly less than 5 mm**
• HOW IT FEEDS **Parasite, sucks blood or body fluid of its host.**
Fleas have long, sharp, hollow, needle-shaped mouth parts to pierce skin – even the thick, leathery hides of hosts such as cattle. Their legless larvae (young) look like very tiny worms or fly maggots.

Fruit fly (p 16)
Drosophila (Diptera)
• SIZE RANGE **Length about 3 mm**
• HOW IT FEEDS **Sucks and laps blobs of juices and soft plant matter.**
Fruit flies gather around ripe or rotting fruit in autumn, and sometimes plague picnics by landing in beer, pickles or other items with bitter or savoury, vinegar-based flavours. They are also bred in billions in laboratories, for genetic research.

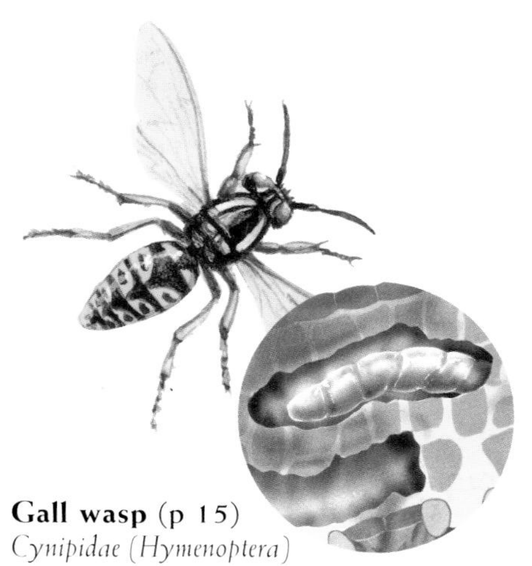

Gall wasp (p 15)
Cynipidae (Hymenoptera)
• SIZE RANGE **Length usually 1–5 mm**
• HOW IT FEEDS **Chews plant matter, some types also catch tiny prey.**
The grubs (larvae) of gall wasps tunnel into parts of plants, such as leaves or stems, eating as they go. The plant reacts by trying to "wall off" the invader by making a hard capsule or container, called the gall.

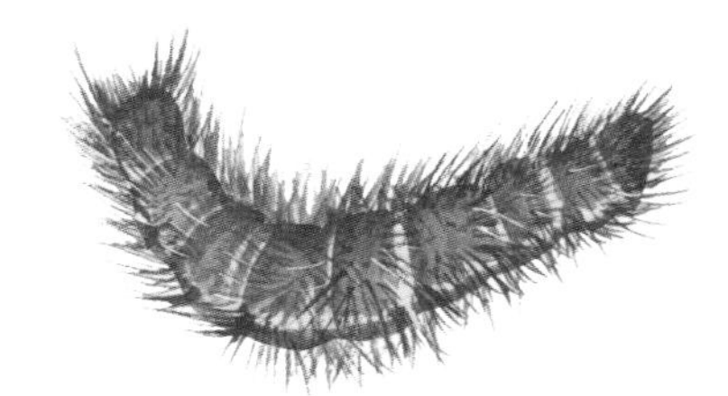

Geometrid moth caterpillar (p 14)
Geometridae (Lepidoptera)
• SIZE RANGE **Up to 2 mm long**
• HOW IT FEEDS **Caterpillar eats leaves and other plant matter.**
Adult geometrid moths are mostly small and brown. It is difficult to distinguish between the 12,000-plus species. Their caterpillars (larvae) are also small and brown, found almost everywhere, yet seldom noticed.

Gnat (p 9)
Culicidae (Diptera)
• SIZE RANGE **Length usually 1–5 mm**
• HOW IT FEEDS **Flies are generalist fluid-feeders, licking and sucking sap, nectar or blood.**
Gnats are small cousins of houseflies and mosquitoes. Some feed on flowers and tiny scraps of plant matter. Others are parasites, biting larger animals and sucking their blood or body fluids, leaving an itchy spot.

Greenfly (p 17)
Macrosiphum and others *(Hemiptera)*
• SIZE RANGE **Length 1–5 mm**
• HOW IT FEEDS **Sucks sap and other plant juices through needle-like mouth parts.**
There are many kinds of aphids, including greenflies and blackflies. Each specializes on a certain plant, such as roses, currant bushes, beans, apple or pear trees. Females breed rapidly without mating (parthenogenesis).

Honeybee (p 16)
Apis (Hymenoptera)
• SIZE RANGE **Adult length 15 mm**
• HOW IT FEEDS **Collects nectar, pollen and other food from flowers.**
Honeybees in the hive "dance" to tell others about the location of nectar-rich blossom. Bees begin their working lives by cleaning and repairing the nest, and only fly out to flowers on feeding trips when they are older.

Housefly (p 32)
Musca and others *(Diptera)*
• SIZE RANGE **Length 5–10 mm**
• HOW IT FEEDS **Adult regurgitates saliva and digestive juices on to food.**
Houseflies do not spread disease by biting, but by carrying bits of rotting food, excrement, manure, decaying flesh and other germ-ridden matter on their bodies. Their legless larvae are called maggots.

Ladybird (p 22)
Coccinella and many others *(Coleoptera)*
• SIZE RANGE **Mostly less than 10 mm**
• HOW IT FEEDS **Preys on greenflies, small caterpillars and similar victims.**
The bright colours of ladybirds, which are small beetles, warn predators that their flesh tastes horrible and so they are to be avoided. There are many species, mainly red, yellow or orange with dark spots or blotches.

Leatherjacket (p 24)
Tipula and others
• SIZE RANGE **20–30 mm**
• HOW IT FEEDS **Eats the roots and other underground parts of various plants.**
Named after their tough skin, leatherjackets are the soil-dwelling larvae (young) of craneflies. They can become major pests of root crops, corms, bulbs and the roots of grasses in pastures and meadows.

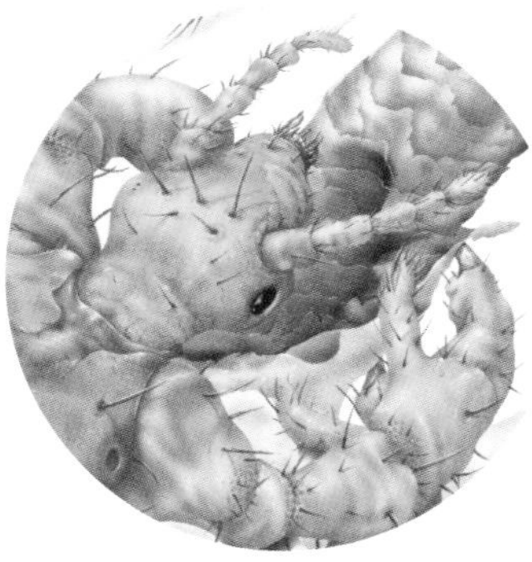

Louse (p 28)
Anopleura
• SIZE RANGE **Less than 5 mm long**
• HOW IT FEEDS **Sucks the blood and body fluids of their hosts.**
Each kind of *anopleuran* or parasitic louse has its preferred host, which in the vast majority of species is a mammal or bird. These lice lack wings and travel by clinging to a convenient passer-by.

Springtail (p 21)
Collembola
• SIZE RANGE **Mostly less than 2 mm**
• HOW IT FEEDS **Chews any small, edible particles, or sucks plant juices.**
Springtails are small, simple, wingless, insect-like creatures that thrive in soil, rotting matter and on various plants. They may multiply to become pests, covering crops such as sugar cane like black dust.

Scale insect (p 31)
Coccidae (Hemiptera)
• SIZE RANGE **Mostly less than 2 mm**
• HOW IT FEEDS **Sucks juices from trees, crops and other plants.**
Scale insects move about freely when nymphs (young forms). The females drive their beak-like mouths into a plant and then attach themselves to it, making a hard, scale- or shell-like covering for the body.

Silverfish (p 33)
Lepisma and others *(Thysanura)*
• SIZE RANGE **Length up to 15 mm**
• HOW IT FEEDS **Munches scraps of food, especially starches in bread.**
This silvery insect comes out to feed at night on tiny bits of food, in larders, kitchens and storerooms. It runs away rapidly into a crevice if the light is switched on. It is a type of three-pronged bristletail.

Tortoiseshell butterfly (p 9)
Nymphalis (Lepidoptera)
• SIZE RANGE **Wingspan 60 mm**
• HOW IT FEEDS **Most butterflies have tube-shaped mouth parts to sip nectar.**
The name of the butterfly-and-moth group, *Lepidoptera*, means "scale-wings". This describes the tiny, soft scales which cover the wings and produce their beautiful colours and patterns.

Woodworm (p 18)
Anobium and many others *(Coleoptera)*
• SIZE RANGE **Length up to 5 mm**
• HOW IT FEEDS **Eats its way through wood, digesting the nutritious parts of the wood.**
Hundreds of kinds of beetle grubs (larvae) bore and tunnel through wood, sometimes for many years. Among the most familiar are the maggot-like larvae of the furniture beetle (domestic woodworm) which attack beams and furniture.

ARACHNIDS

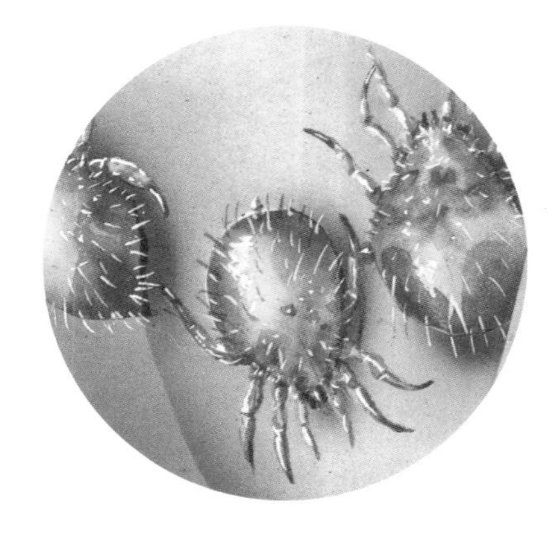

Dust mite (page 30)
Astigmata (Acari)
• SIZE RANGE **Mostly less than 1 mm**
• HOW IT FEEDS **Scavenges on bits of skin, hair, plant fragments.**
Various kinds of dust mites eke out a living in houses and other buildings. They feed on unseen bits of natural debris from people, pets, food, houseplants and other sources.

False scorpion (p 21)
Pseudoscorpiones
• SIZE RANGE **Mostly 2–8 mm**
• HOW IT FEEDS **Preys on tiny animals in soil and among leaves.**
False scorpions are close cousins of scorpions, with eight legs and two pincers, but they lack the tail sting. They sometimes grab on to passing animals, such as beetles, flies or birds, to hitch-hike a lift to a new area.

Flower spider (p 17)
Misumena and many others *(Araneae)*
• SIZE RANGE **Body length 1–10 mm**
• HOW IT FEEDS **Waits in or near a flower to bite and poison small prey.**
Flower spiders such as the crab, small-wolf and petal spiders use blooms as hunting grounds. They hide behind the petals, or even camouflage themselves the same colour as the flower, to ambush visitors.

Orobatid mite (p 22)
Orobatidae (Acari)
• SIZE RANGE **Less than 1 mm**
• HOW IT FEEDS **Chews tiny fragments of anything edible.**
Soil mites tend to have tougher body casings than other mites, as protection against being squashed between mineral particles, hence the common name of beetle-mites. There are predatory, herbivorous and scavenging types.

Proturan (p 22–23)
Protura (Insecta)
• SIZE RANGE **Up to 4 mm**
• HOW IT FEEDS **Pierces and sucks up any nutritious liquids.**
Proturans are found almost everywhere in soil, leaf litter and rotting plant matter. They are tiny, pale and wingless. Their long front legs feel and detect tastes, while the rear two pairs are used for movement.

Red velvet mite (p 19)
Trombididae (Acari)
• SIZE RANGE **Mostly up to 2 mm long**
• HOW IT FEEDS **Chews tiny edible fragments of any kind.**
Many mites are parasites of larger animals. But most kinds of red velvet mites are free-living, as they wander in search of any food. They are seen on wooden window frames and doors in sunny weather.

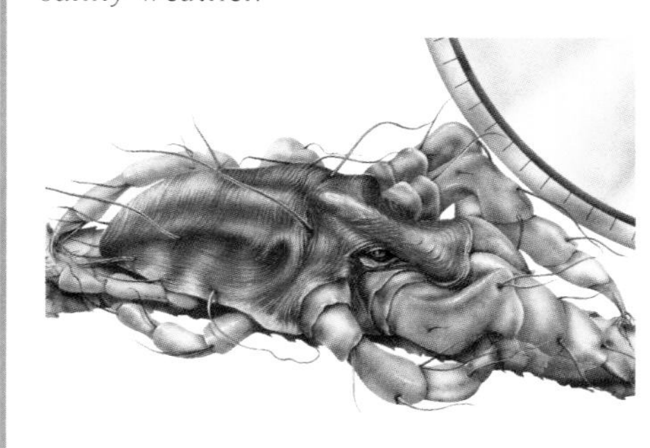

Skin mite (p 29)
Sarcoptes (Acari)
• SIZE RANGE **Length 1–3 mm**
• HOW IT FEEDS **Burrows in skin, feeds on skin and other tissues.**
Various kinds of mites attach to people and feed on skin flakes or the soft tissue beneath. The scabies mite causes intense, painful itching as it burrows.

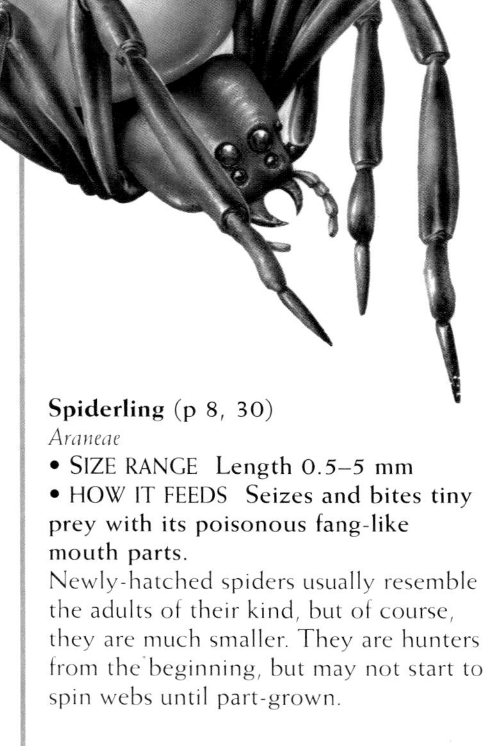

Spiderling (p 8, 30)
Araneae
• SIZE RANGE **Length 0.5–5 mm**
• HOW IT FEEDS **Seizes and bites tiny prey with its poisonous fang-like mouth parts.**
Newly-hatched spiders usually resemble the adults of their kind, but of course, they are much smaller. They are hunters from the beginning, but may not start to spin webs until part-grown.

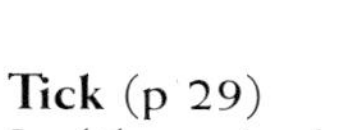

Tick (p 29)
Ixodidae and others *(Acari)*
• SIZE RANGE **Mostly less than 5 mm**
• HOW IT FEEDS **Sucks blood through its strong, beak-like mouth.**
Like many tiny, blood-sucking parasites, there are different ticks for different hosts, such as sheep, cattle, dogs and chickens. However most types can make do and survive on a different host, for a limited time.

CRUSTACEANS

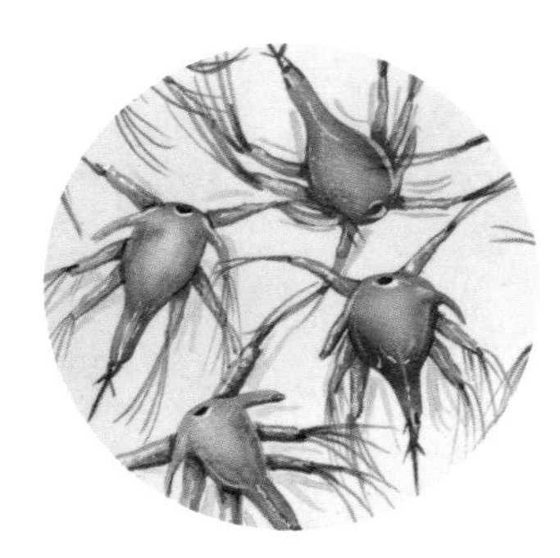

Barnacle (larvae) (p 11)
Balanus (Crustacea)
• SIZE RANGE **Diameter up to 10 mm**
• HOW IT FEEDS **Adult waves its feather-like limbs to gather tiny edible particles from sea water.**
Larvae (young forms) of barnacles drift in the water and change shape many times, before settling on to rocks. They grow into cone-shaped adults, with hard, stony plates covering the soft inner body.

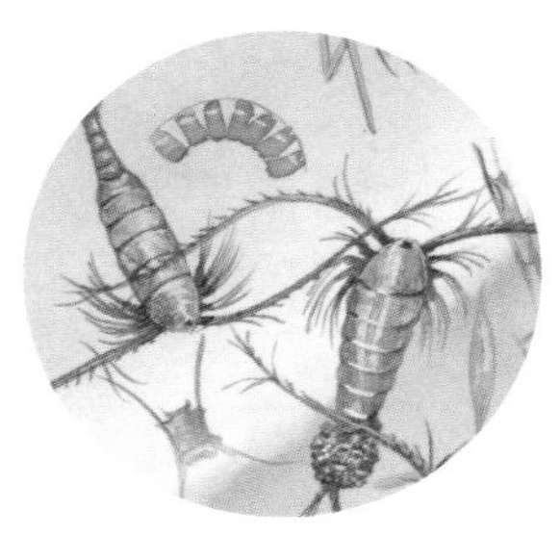

Copepod (p 10–11)
Copepoda (Crustacea)
• SIZE RANGE **Up to 5 mm long**
• HOW IT FEEDS **Filters tiny organisms and food particles with its bristly front legs.**
Most copepods are about 1–5 mm long and occur in gigantic shoals or swarms. They are one of the main links in the ocean food chains between the truly microscopic plankton, and larger animals like fish.

Water flea (p 12–13)
Daphnia and other types
• SIZE RANGE **Length usually 1–3 mm**
• HOW IT FEEDS **Filters tiny particles from water using its large antennae and bristly legs.**
Water fleas swim in a jerky fashion by rowing with their many-branched antennae. In good conditions they form large swarms in lakes and rivers. Some are bred to be sold commercially as fish food.

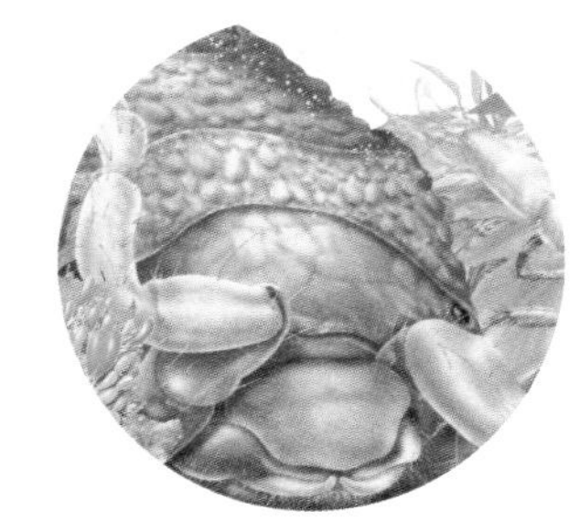

Woodlouse (p 20–21)
Isopoda (Peracarida)
• SIZE RANGE **Length up to 2 cm**
• HOW IT FEEDS **Eats soft, decaying plant matter, fungi and other scraps.**
Also called sowbugs, woodlice gather in damp, dark places such as under bark, logs and stones, and in leaf litter. Most have seven pairs of legs. Some types, known as pillbugs, can roll up into hard-cased balls.

MYRIAPODS

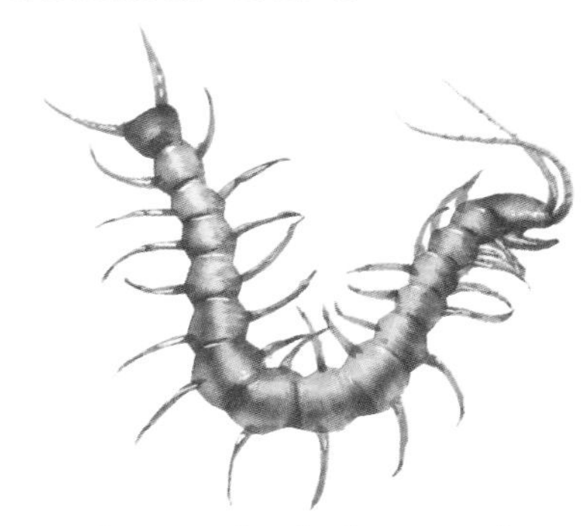

Burrowing centipede (p 24–25)
Halophilus, Necrophloephagus and others
• SIZE RANGE **Length up to 50 mm**
• HOW IT FEEDS **Wriggles through soil, preying on any tiny animals.**
The centipede's first pair of limbs resemble large, sharp fangs and inject poison into the victims they bite. Most centipedes are swift-running predators; soil and burrowing types tend to be slimmer.

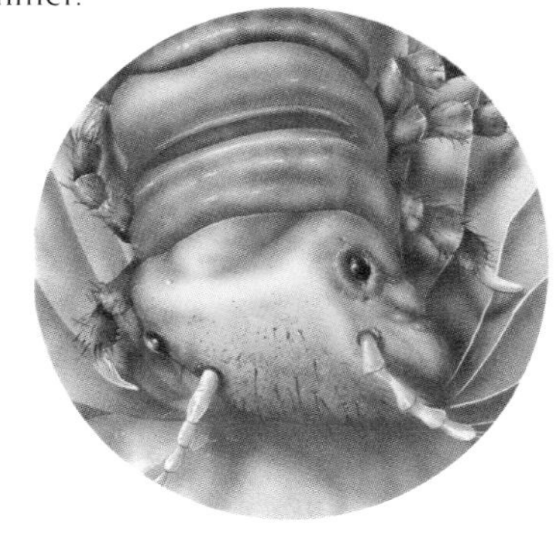

Millipede (p 20)
Diplopoda
• SIZE RANGE **Length up to 5 cm**
• HOW IT FEEDS **Chews old bits of wood, twigs, flowers, fruits and bark.**
Millipedes are slow-moving and shy of light, warmth and dryness. They gather in dark, damp, cool places. Most have foul-tasting or noxious chemicals which they ooze from glands along the body, if they are attacked.

MOLLUSCS

Burrowing slug (p 24–25)
Testacella and others *(Gastropoda)*
• SIZE RANGE **Length up to 12 cm**
• HOW IT FEEDS **Some types eat plant matter, others are predatory.**
Slugs are cousins of snails and move in the same way, gliding on trails of slime. One type of burrower, the shield slug, is a slow but active predator of small worms and similar soil animals – including other slugs.

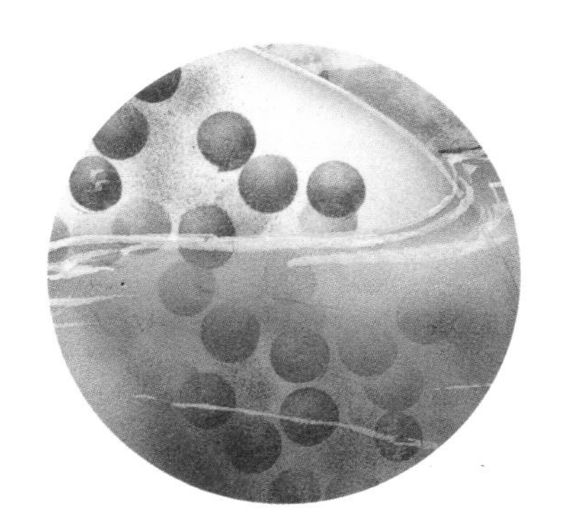

Pond snail (eggs) (p 26)
Lymnaea and others *(Gastropoda)*
• SIZE RANGE **Diameter up to 0.2 mm**
• HOW IT FEEDS **The animal rasps up pieces of waterweeds, decaying matter, and small animals.**
The various kinds of pond snails are mostly vegetarian, but the great pond snail scavenges on decaying fish, and may consume the eggs of fish, insects and other snails.

FISH

Fish fry (p 10–11)
Pisces (overall fish group)
• SIZE RANGE **Length 1–10 mm**
• HOW IT FEEDS **Fry are nourished by the yolk store in their eggs during the early stages of life.**
Fry are the second stage in the life of most fish, after developing from the eggs. However some fish, like various sharks, develop further in their egg cases, and hatch to resemble smaller versions of the adults.

BIRDS

Treecreeper (p 18)
Certhia familiaris
• SIZE RANGE **Beak to tail 12 cm**
• HOW IT FEEDS **Pecks at bark, stems and leaves for small animals.**
The treecreeper has a long, sharp, down-curved bill, to probe into and under bark. It slowly hops up a tree trunk, looking and listening and pecking as it goes, then flies to the base of the next trunk.

Index